solid-STATE circuits guidebook

solid-STATE CIRCUITS guidebook

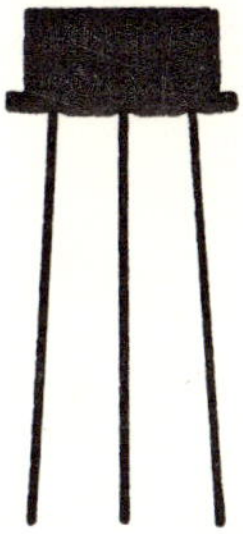

by BRICE WARD

TAB BOOKS
Blue Ridge Summit, Pa. 17214

FIRST EDITION

FIRST PRINTING— JANUARY 1974

Printed in the United States
of America

Hardbound Edition: International Standard Book No. 0-8306-3699-4
Paperbound Edition: International Standard Book No. 0-8306-2699-9

Library of Congress Card Number: 73-86769

PREFACE

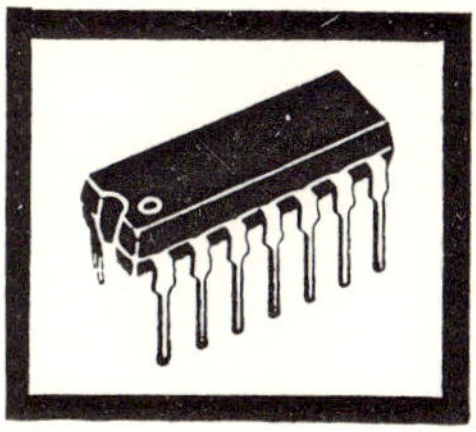

It might be quicker to tell you first what this book is **not**. It's not a troubleshooting book, nor a collection of ready-to-build projects. Neither is it an audio amplifier book, although many modern audio amplifier circuits are included, with a wide variety of power output and load-driving capabilities.

The book is an **idea stimulator**. It discusses many circuits, each one very useful and practical—and each designed to stimulate **circuit design ideas**. Each circuit description is keyed to a schematic, complete with parts identifications and values.

Anyone interested in and familiar with semiconductors, and who is reasonably facile with mathematics, will find these circuits and ideas useful for **experiments**, for **teaching**, for **actual projects**—for any purpose, limited only by the reader's imagination—all leading to a deeper understanding of solid-state electronic design. Not only can these circuits be modified and used for purposes other than those stated, they can be strung together to perform multiple or different functions. Again, **imagination**!

Part I covers a wide range of circuits using discrete components: amplifiers, multivibrators, counters, digital display circuits and digital-to-analog converters, ramp and staircase wave generators, pulse counters, Schmitt triggers, complementary circuits, diode transistor logic gates, regulated power supplies, and a variety of miscellaneous circuits. Part II is devoted to the design and construction of a wide variety of audio power amplifiers from 1 to 100 watts output, with circuit card layouts, parts lists, and performance specifications. Part III describes several integrated circuits using generally available linear ICs.

To save you many hours of eyestrain and paperwork, all transistors, diodes, and other semiconductors called for in the schematics by Ferranti identifications or by standard designations have been tabulated numerically and cross referenced to Pro Electron, RCA, Motorola, Sylvania, and General Electric type numbers. This information was extracted from the manufacturers' published replacement cross-reference guides, as normally available at distributors' counters. As always, bear in mind that there may be variations in characteristics of components from one manufacturer to another. At worst, minor modification may be needed.

We gratefully acknowledge the assistance of Ferranti, Ltd., Gem Mill, Chadderton, Oldham, England, and thank them for permission to use material from their E-Line Transistor Applications Booklet. This material, containing characteristic curves, specifications—design data sheets, is in the appendix. Our thanks also to Motorola Semiconductors, Ltd., York House, Wembley, Middlesex, England, for supplying additional information and assistance.

Special thanks to Mr. Ira S. Gewant of Ferranti Electric, Inc., E. Bethpage Road, Plainview (L.I.) N.Y. 11803, for researching much data and supplying cross references for several discrete components for which U.S.A. equivalents were not readily available.

Waveform photographs were taken with a Polaroid Corp. Land camera from the faceplate of an Advance Electronics model OS1000 oscilloscope. We thank these firms for their cooperation and assistance.

Brice Ward

TABLE OF CONTENTS

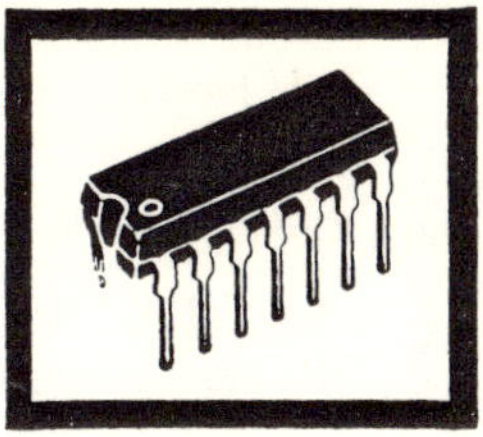

Part I Solid-State Circuits Using Discrete Components

Part I

Solid-State Circuits Using Discrete Components

Introduction, and Special Cross-Reference

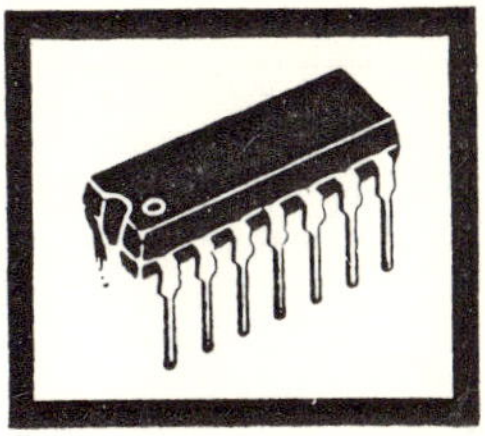

Part I, which is intended for students, amateurs, and engineers, provides a basic working knowledge of semiconductor characteristics, and aids those who are unfamiliar with transistor switching characteristics to become familiar with them through building experimental circuits. Usually, the design procedures given herein were tailored for worst-case parameters; this will normally result in a circuit which cannot fail to operate. However, your requirements will often differ in some details from the examples given and the conditions present at the input and output of the circuit will be unique to your own system. Thus, in most cases, it will be necessary for you to design a circuit to your own specifications. The examples given will assist in this, but the finished circuit should always be tested in breadboard form, preferably with worst-case component values, to ensure that its operation is satisfactory.

As stated in the preface, many devices listed in schematics herein are identified by Ferranti or British standard numbers. For convenience, since you, the reader, will be constantly referring to it—Table A cross references the Ferranti E-Line numbers to U.S.A. equivalents.

Table A. Ferranti to U.S.A. Semiconductor Type Cross References

FERRANTI NO.	PRO-ELECTRON (BRITISH STD)	MOTOROLA MPS—	SEMICONDUCTOR HEP—	RCA SK PREFIX	SYLVANIA ECG PREFIX	GENERAL ELECTRIC GE—	SPECIAL NOTES OR U.S. JEDEC NO.
DIODES							
ZTX531							npn planar: GP: plastic encapsulated
BFS59	BFS59	VCEO 25-45					
KS34B							Zener. 4.3V 1N5229B
KS043A							Zener. 5.1V 1N5231B
KS34B	B.S.9300-C540						Zener. 00.1V: 1N962B
KS110A	CV7144						50V RWM: 250 mA at 25 deg C, rectifier
ZS120			156	3030	116	504A	15VRWM: Ultrahigh-speed switching
ZS140	B.S.9300-C730						B4A5; Edal Industries, E. Haven, Conn.
ZS170	B.S.9330-F007 (Q)						100 PIV: DO-41 JEDEC: rectifier
ZS271	B.S.9330-F014 (Q)						
HIGH SPEED							
ZDX1R	PLANAR PLASTIC ENCAPSULATED DIODE PAIR—300 mW max at 25 deg C; forward I per common lead, 250 mA; $15V_{RWM}$; forward drop 1V at 10 mA; Reverse recovery time of 7 usec when $I_F = I_R = 50$ to 400 mA; capacitance (max) 4 pF; Reverse I, 1 uA. Common anode						
TRANSISTORS	(SEE APPENDIX FOR CURVES AND SPECS)						
PREFIX ZTX							
107	BC107; BC182L	3694	56	3018	123A	11	General purpose npn
108	BC108; BC183L	A20				62	" " "
109		A20				62	" " "
114		6532	S0004	3039	107	11	Low-noise, high-gain npn
300	BCW10	A20				20	General purpose npn. Complement: **ZTX500**
301	BCW12	A20				20	" " " " ZTX501
302	BCW14	A20				20	" " " " ZTX502

303	BCW16	6560	S0002	3122	123A	63	" " " " ZTX503
304	BCW18	A06	53	3024	128	18	" " " " ZTX504
310	BSV23: CV9211	6514	728	3018	123A	61	ZN706: high-speed switching npn.
311	BSV24: B.S. 9300-C464	6514	728	3018	123A	61	2N706A: high-speed switching npn.
312	BSV25					20	2N2369: " " " "
313	BSV26: B.S.9300-C555						2N2369A: " " " "
320	BFW97	918	720	3039	107	11	2N918: VHF—UHF npn
321		918	720	3039	107	11	VHF—UHF npn
325		H10					UHF npn
326		H10					UHF npn
330	BCW20		729	3124	123A	20	2N3707, 2N5209
331	BCW22		729	3124	123A	20	2N5209: low-noise, low-level npn
341	BSV28	918	720	3039	107	11	Nixie driver npn
342	BSV29	918	720	3039	107	11	" " "
500	BCW11	A70		3114	159	67	General purpose pnp. Complement: ZTX300
501	BCW13	A70		3114	159	67	" " " " ZTX301
502	BCW15	A70		3114	159	67	" " " " ZTX302
503	BCW17 (2N5086)		715	3025	127	22	" " " " ZTX303
504	BCW19	A56					" " " " ZTX304
510	BSV33		52	3004	102A	53	2N2894: 2N5028: TIS50: pnp
530	BCW21	A70		3114	159	67	pnp Complement: ZTX330
531	BCW23		715	3025	127	22	2N5086, pnp. Complement: ZTX331
BFS59	BFS59 V_{CEO} 25-45: 500 mW at 25 deg. C.. Min. f_T over 150 MHz. min h_{FE} 35-50: I_C 500 mA peak npn planar: GP: plastic encapsulate: V_{CBO} 60V: V_{CEO} 30V: V_{EBO} 5V: 0.5W at 25 deg C Ambient. V_{CE} (sat) 0.35V at IC 150 mA: h_{FE} 40-300 at IC 150 mA. I_{CBO} 0.1 uA at 25 deg C at 150 MHz f_T						
LOGIC							
ZN222E	9 msec molded dual in-line 0 to 70 deg C						DTL IC: dual JK flip-flop
ZN322E	15 msec molded dual in-line 0 to 70 deg C						DTL IC: dual JK flip-flop
ZN7474E	10 msec molded dual in-line 0 to 70 deg C						Any 7474 dual D-type flip-flop

Chapter 1

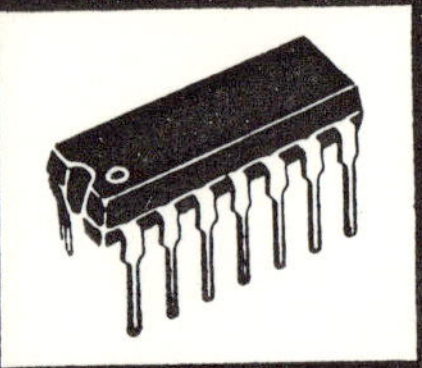

Switching Characteristics of Transistors

In most switching circuits, trigger capacitors are required to differentiate pulses and to provide dc isolation between stages, and speedup capacitors are required to neutralize the stored charge and junction capacitances of the transistors. It usually suffices to choose values small enough not to interfere with the operation of the circuit at its maximum frequency. These capacitors, however, do have minimum values depending on the type of transistor used, its collector and base currents, and the amplitudes of the waveforms involved.

In order to understand the need for speedup capacitors, consider the simple inverter circuit of Fig. 1-1.

Initially, npn transistor Tr1 is cut off, with its base-emitter junction reverse biased to —1.67V and its collector-base junction reverse biased by 6.67V. When the input pulse arrives, the potential at point A rises exponentially as the transistor's junction capacitances charge, until point A reaches 0.7V and the base-emitter junction of the transistor starts to conduct. The time taken for this to occur is called the delay time (t_d). As the base current flows, charge carriers begin to accumulate in the base region and the collector current increases in proportion until it is limited by the 1K collector load resistance, R3. The time taken for the collector current to increase from 10 percent of its final value to 90 percent of its final value is called the rise time (t_r). As the collector voltage approaches zero, the collector-base junction becomes forward biased and conducts the excess base current. The base now starts acting as an emitter and begins to inject charge carriers into the collector. This causes a considerable accumulation of charge in the collector region which must be removed, either by recombination or by reverse base current, before the transistor can begin to turn off. The time taken for this excess charge to be removed is called the storage time (t_s), and during this time the transistor remains saturated. At the end of the storage time, all excess charge carriers have been removed and the base charge is simply that required to maintain collector current. As this charge is reduced further, the collector current falls in sympathy until the transistor is cut off. The time taken for the collector current to fall from 90 percent of its previous value to 10 percent of its previous value is called the fall time (t_f).

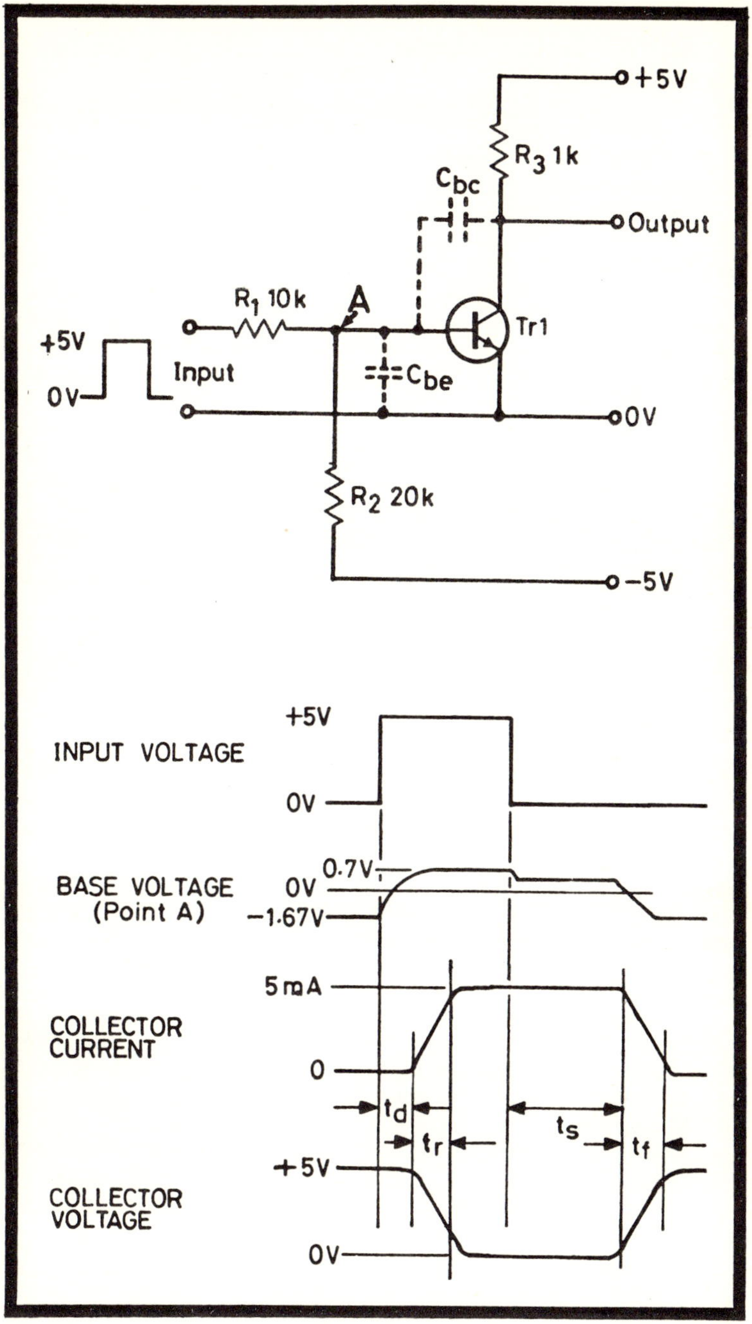

Fig. 1-1. Simple transistor inverter, with switching waveforms.

Because of the low leakage currents of silicon transistors, it is not usually necessary to reverse bias the base to ensure dc stability. This reduces delay time, but causes a large increase in both the storage time and fall time, because the charge can only be dissipated by recombination instead of by the reverse base current provided by R2. This situation can be overcome by connecting a capacitor in parallel with R1 so that the charge stored on the speedup capacitor is sufficient to neutralize the total charge in the conducting transistor, i.e., the base charge, the collector storage charge, and the junction capacitances. Figure 1-2 shows the effect of varying speedup capacitor C from zero to a critical value at which all the charge is just removed. This value depends on three factors: (1) transistor type, (2) base and collector currents, and (3) amplitude of the input pulse. As no stored-charge information is available for the Ferranti E-Line transistors, nor for Motorola equivalents, it is necessary for the minimum values to be determined experimentally, preferably in a worst-cases version of the actual circuit. However, as a guide, the minimum values of charge necessary to turn off devices of the ZTX300/500 (MPS-A20/A70), ZTX310 (MPS6514), ZTX510 (2N5208), and ZTX107 (MPS3694) series have been plotted for various dc conditions in Fig. 1-3. The circuit used for measuring the minimum values of trigger charge and the results obtained are shown in Figs. 1-4 and 1-5. The values obtained are higher than the above case because the trigger capacitor not only has to remove all the stored base charge, but also has to overcome the dc bias provided by Rb.

In a practical circuit, there will usually be steering diodes and resistors or other circuit components, which will absorb a proportion of the trigger charge. Consequently, it will usually be necessary to use a higher value of trigger capacitance than is suggested by the curves, which should be treated only as a guide. It is always advisable to measure the values under working conditions in a breadboard circuit built with worst-case component values.

Only the highest figure has been recorded for ZTX310, as differences were small.

It should be pointed out that these measurements were taken from the worst samples out of a batch of 50 and, as these parameters are not tested during manufacture, no guarantee can be given that future production will conform to these results.

The transient behavior of transistors is a complex subject, and this summary has, of necessity, been greatly simplified. Motorola's "High Speed Switching Transistor Handbook"* is suggested for further reading.

---

*Technical Information Center, Motorola Semiconductor Products, Inc., P.O. Box 20912, Phoenix, Arizona 85034.

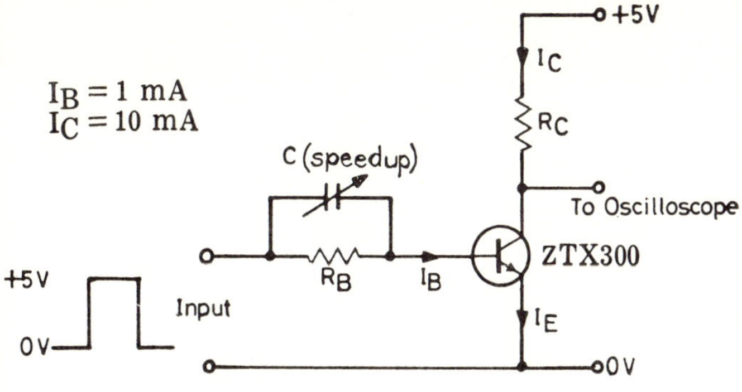

Upper trace—Input waveform

Lower traces—(a) C = 0
(b) C = 100 pF
(c) C = 120 pF
(d) C = 130 pF (stored charge just removed)

Note the reduction in delay time from trace (a) to traces (b), (c), (d).

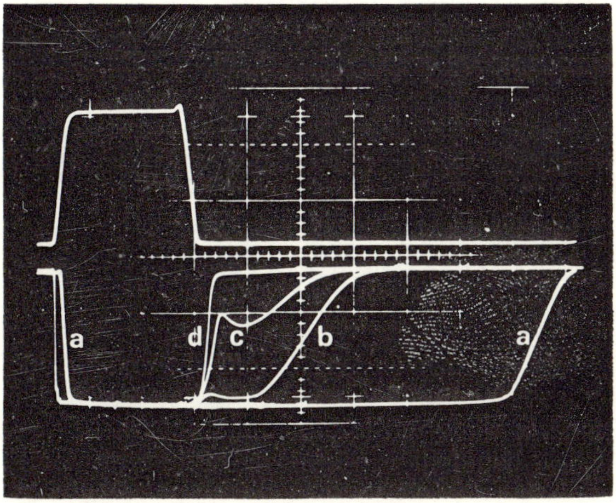

Horizontal scale: 0.2 usec per cm
Vertical scale: 2V per cm

Fig. 1-2. Circuit for measuring stored charge, with waveforms.

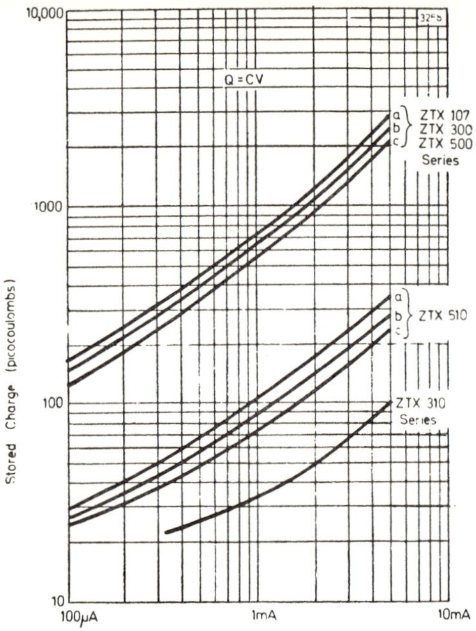

Curves marked (a) $I_C/I_B = 10$
(b) $I_C/I_B = 20$
(c) $I_C/I_B = 40$

There was no significant difference for ZTX310 series transistors.

Fig. 1-3. Stored charge vs base current for various transistors.

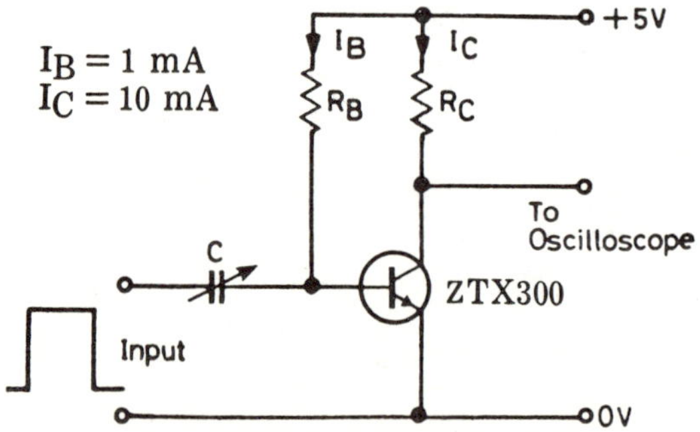

Fig. 1-4(A). Circuit for measuring minimum trigger charge, with waveforms. (Scope settings same as in Fig. 1-2).

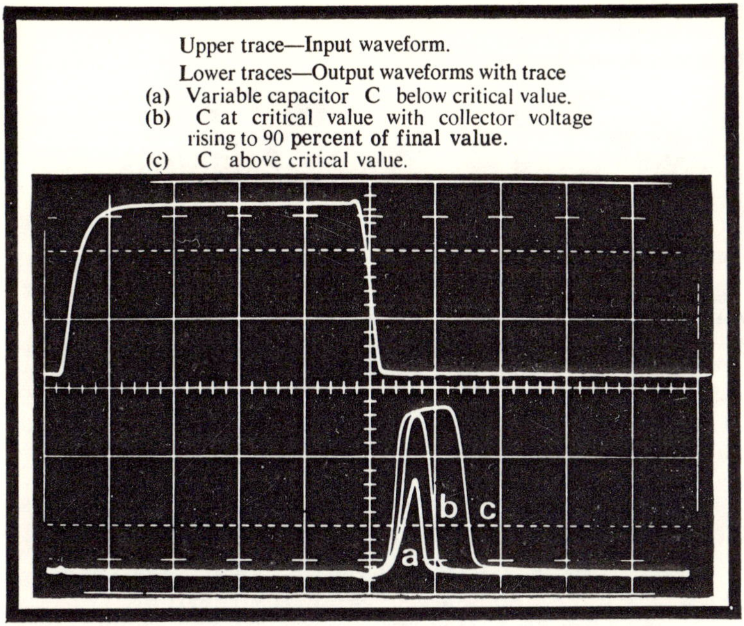

Fig. 1-4B. Circuit for measuring minimum trigger charge, with waveforms. (Scope settings same as in Fig. 1-2).

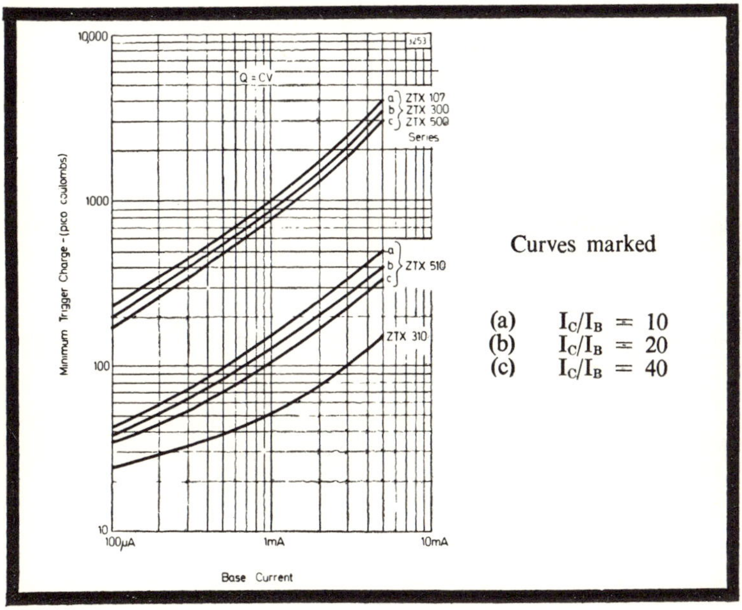

Fig. 1-5. Minimum trigger charge vs base current for same devices as in Fig. 1-3.

Chapter 2

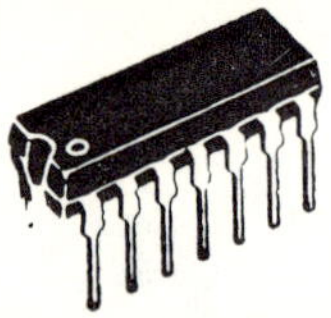

Calculating for Simple Amplifiers

Figure 2-1 shows two methods of biasing a simple transistor amplifier stage. The method shown in Fig. 2-1A is widely used with germanium transistors, and is necessary because of their significant collector-base leakage current. In silicon transistors, Icbo is usually negligible and the configuration of Fig. 2-1B is adequate for most applications, with a saving of two resistors and one electrolytic capacitor.

Usually the circuit is designed so that the collector voltage Vce is approximately half the supply voltage Vcc. This is achieved when:

$$R_b = h_{FE} \times R_c$$

where hFE is the direct-current gain of the transistor.

In practice, a wide range of current gains are encountered;
e.g., ZTX300 (MPS-A20) h_{FE} 50 to 300 at 10 mA collector current, so a value of h_{FE} given by:

$$h_{FE} = \sqrt{h_{FE(max)} \times h_{FE(min)}}$$

is used. For the MPS-A20, $h_{FE} = \sqrt{50 \times 300} = 122.5$. (We will use this number later.)

The choice of R_c depends upon the supply voltage and the amplification of the stage. For low-level amplification, a low value of collector current (50 to 300 uA) is necessary to minimize noise and R_c will have a high value, of the order of 47K. For wideband amplification, a low value of R_c is necessary to minimize the effect of stray capacitance, and a correspondingly high collector current (e.g., 10 mA) is used to maximize cutoff frequency f_T.

EXAMPLE OF DC DESIGN FOR A SINGLE-STAGE AMPLIFIER

For the following work, refer to Fig. 2-2.

$$I_c = h_{FE} \times I_b$$

$$I_e \text{ (the current through } R_c) = I_c + I_b = I_b(h_{fe} + 1)$$

$$\approx h_{FE} \times I_b$$

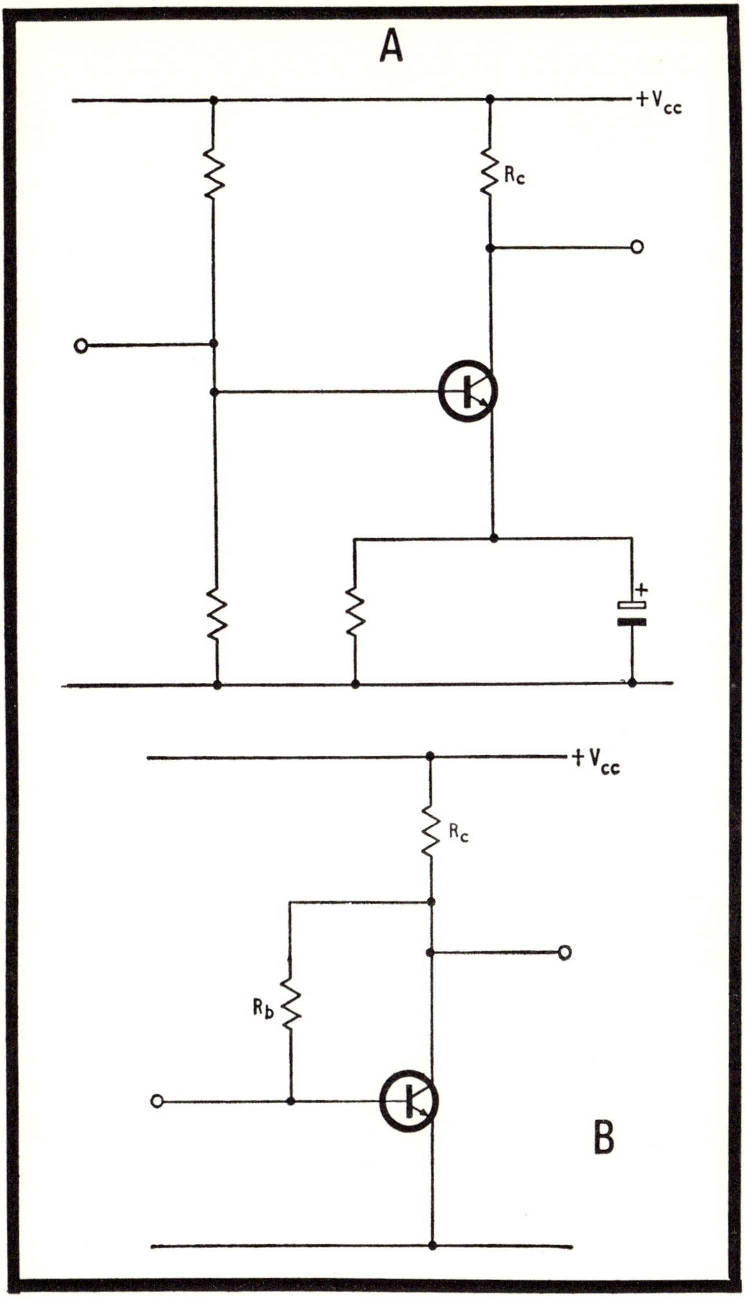

Fig. 2-1. In A, generalized transistor biasing scheme. In B, biasing scheme for germanium transistors in particular.

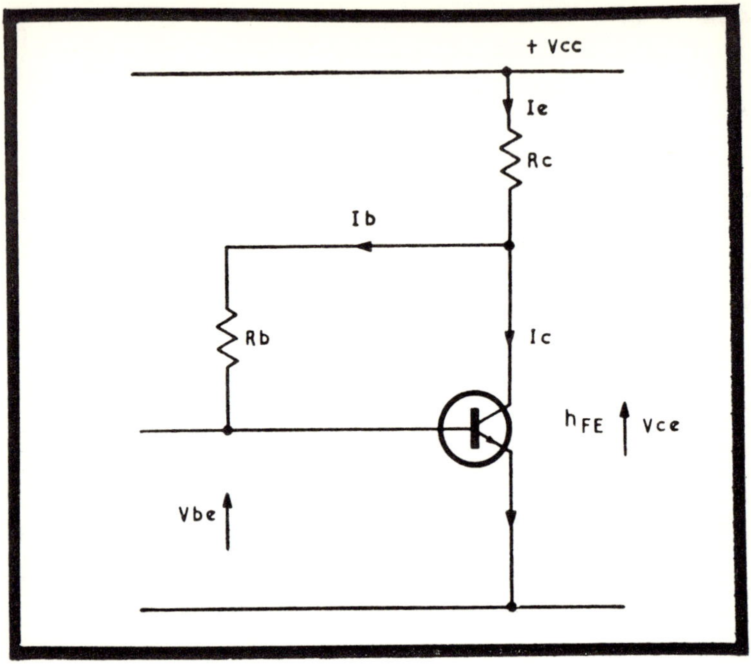

Fig. 2-2. Single-stage amplifier.

since $h_{FE} \gg 1$

Therefore, $I_e \approx I_c = h_{FE} \times I_b$

$$I_b = \frac{V_{ce} - V_{be}}{R_b} \mathrel{\dot{\approx}} \frac{V_{cc} - h_{FE} I_b R_c - V_{be}}{R_b}$$

$$= \frac{V_{cc}}{R_b} - \frac{h_{FE} I_b R_c}{R_b} - \frac{V_{be}}{R_b}$$

Therefore, $I_b = \left[1 + \frac{h_{FE} R_c}{R_b} \right] = \frac{V_{cc} - V_{be}}{R_b}$

$$I_b = \frac{V_{cc} - V_{be}}{\frac{R \quad (1 + hFE\ Rc)}{Rb}} = \frac{V_{cc} - V_{be}}{\frac{V_{cc} - V_{be}}{R_b + h_{FE} R_c}}$$

and it follows that

$$I_c = h_{FE} \quad I_b = \frac{h_{FE}(V_{cc} - V_{be})}{R_b + h_{FE} R_c}$$

and $V_{ce} = V_{cc} - I_c R_c$.

Let Vcc 10V and Rc 1K, and using hFE 122.5 (previously calculated) for the ZTX300 (MPS-A20), we can see that $R_b = 122.5$ K 120K preferred, using $h_{FE} \times R_c$.

Further, to find the collector voltage, we use

$$I_e = \frac{h_{FE}(V_{cc} - V_{be})}{R_b + h_{FE} R}$$

$$= \frac{122.5 \quad (10 - 0.7)}{120 + (122.5 \times 1)} \text{—mA}$$

$$= \frac{122.5 \times 9.3}{242.5} = 4.7 \text{ mA}$$

$= 10 - (4.7 \times 1)$ 5.3V nominal collector voltage.

Of course, these calculations would be for a wideband amplifier.

In practice, the current gain of the transistor may lie between 50 and 300 and the resistor values may be, say, 20 percent higher or lower than their nominal values.

There are two worst-case conditions (assuming constant supply voltage).

Case 1

Where h_{FE} is a minimum (50), R_c is a minimum (800 ohms), R_b is a maximum (144K).

$$I_c = \frac{50 \ (10 - 0.7)}{144 + (50 \times 0.8)}$$

$$= \frac{50 \times 9.3}{188} = 2.47 \text{ mA}$$

$V_{ce} = 10 - (2.27 \times 0.8) = 8$V.

Case 2

Where h_{FE} is a maximum (300), R_c is a maximum (1.2K), R_b is a minimum (96K).

$$I_c = \frac{300 \times 9.3}{96 + (300 \times 1.2)}$$

$$= \frac{300 \times 9.3}{456} = 6.14 \text{ mA}$$

$V_{ce} = 10 - (6.14 \times 1.2) = 2.64$V

In Case 1, only 2V is dropped across the load resistance, giving a maximum output of 4V p-p In Case 2, the collector voltage can only fall by 2V before the transistor saturates, so the maximum output swing is again only 4V.

For comparison, an output of nearly 9V p-p could be obtained with all values at the design centers. The possible variation in collector voltage could be minimized by using

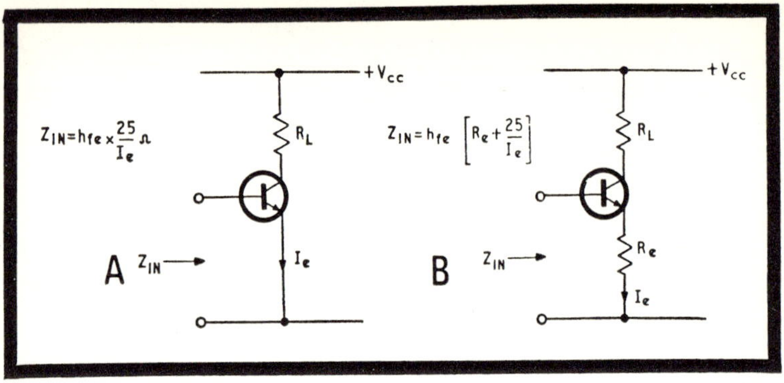

Fig. 2-3. Input impedance, A; in B, input impedance with emitter resistor.

close-tolerance resistors and a narrow gain-spread transistor, such as the ZTX302 (MPS-A20).

INPUT IMPEDANCE

The input impedance of a transistor connected in the common-emitter configuration is given approximately by h_{fe} x r_e where r_e is the intrinsic emitter resistance and is approximately equal to

$$\frac{25}{I_e}.$$

With reference to Fig. 2-3A, I_e is the emitter current in mA, and r_e is the emitter resistance in ohms.

When an external emitter resistor (R_c) is added, the input impedance becomes h_{FE} (r_e x R_e), as in Fig. 2-3B.

Voltage Gain

Voltage gain=current gain x impedance gain =

$$\approx \frac{h_{fe} \times RL}{h_{fe} \times r_e} \approx \frac{(RL)}{r_e} + \frac{RL}{r_e + R_e}$$

i.e., voltage gain is independent of h_{fe}.

Feedback

In the simple bias circuit discussed earlier, a small amount of voltage-derived feedback is present (i.e., a feedback current is produced proportional to the output voltage because of the bias resistor between base and collector). This feedback reduces both the input impedance and the overall current gain in the same ratio. The factor by which they are reduced is given by:

$$\frac{1}{1 + h_{fe}\frac{R_L}{R_b}}$$

where h_{fe} = transistor ac gain
R_b = bias resistor
R_L = effective ac load resistance, i.e., collector resistor R_c in parallel with the input impedance of the next stage. See Fig. 2-4.

Current Feedback

When a resistance R_e is put in series with the emitter, a voltage is developed across it which is proportional to the emitter (and hence collector) current. Current feedback increases the input impedance, but does not affect the current gain of the stage.

WIDEBAND LINEAR RF AMPLIFIER

The ZTX320/BFW97 (MPS918) rf transistor can be used in a two-transistor feedback pair circuit to provide a current amplifier of controlled gain and bandwidth. The circuit is shown in Fig. 2-5.

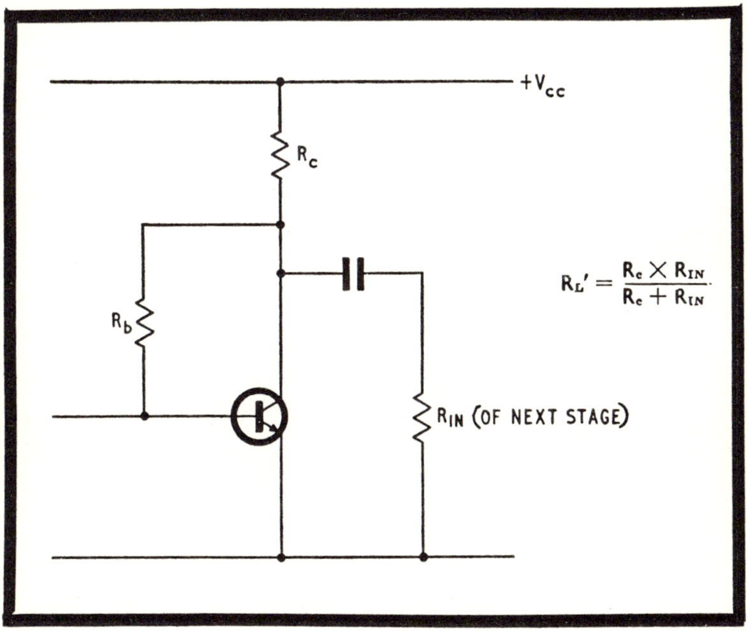

Fig. 2-4. Feedback and R_L.

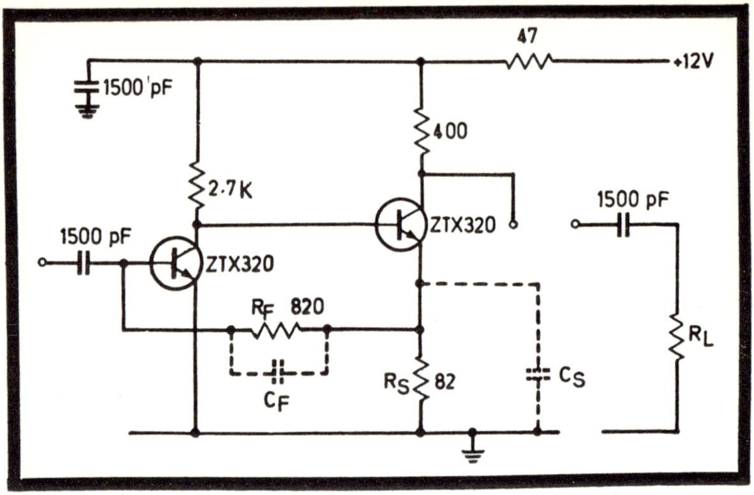

Fig. 2-5. Current amplifier with feedback.

The current gain is given by the resistor ratio

$$\frac{R_F + R_S}{R_F} = 10$$

The circuit exhibits a peak in the high-frequency response curve, caused by stray capacitances across the emitter and collector resistors. The addition of compensating capacitors C_F and C_S will give a flat frequency responce. Owing to the high frequencies involved, the components should be assembled closely together to reduce stray capacitance and lead inductance.

Typical performance for the circuit is:

Gain = 10
Bandwidth at 3 dB down = 230 MHz
Current from 12V supply = 12 mA
Compensation capacitors: CF=1 pF; CS=2 pF

The nature of the feedback gives a very low input impedance, typically less than 10 ohms; the circuit was fed by the matching network shown in Fig. 2-6.

The 390 ohm feed resistor gives an input resistance equal to the output resistance, and a comparison of input voltage across the 56 ohm shunt resistor and the output voltages can be taken as a direct measure of current gain.

The amplifier power consumption is typically 150 mW at 12V. The quiescent currents of the transistors in Fig. 2-5 are 3 mA and 10 mA, respectively. With a nominal 400 ohm

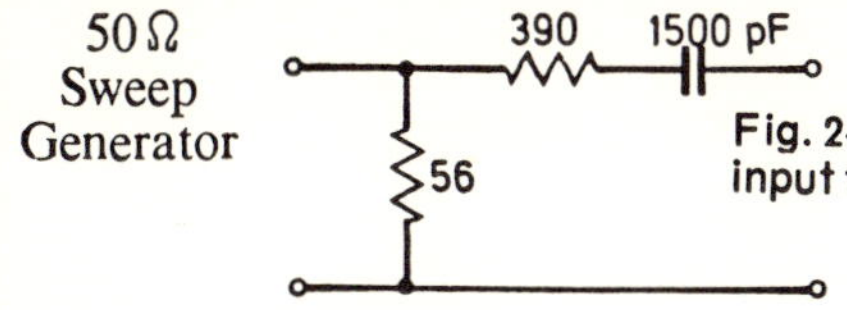

Fig. 2-6. Matching network for input to current amplifier.

collector load, the output voltage is 8V dc quiescent with a 12V supply.

The maximum undistorted output is approximately 7V p-p with no load connected, falling to 1.5V p-p across a 50 ohm load.

Input Matching Stage

A grounded base ZTX320 can be used to provide an input of 50 ohms to match standard 50 ohm cables or sources. The input resistance of the grounded base transistor stage is given by

$$R_{IN} \triangleq r_e \triangleq \frac{25}{I_E\,(\text{mA})}$$

By biasing the transistor to 0.5 mA, a reasonable match can be achieved. The complete circuit is shown in Fig. 2-7.

Typical performance was measured as:
Overall gain (50 ohm source, 50 ohm load) = 9.2, and 3 dB bandwidth = 250 MHz.

The wide bandwidth of the basic circuit shown in Fig. 2-5 can be sacrificed, in exchange for increased current gain, by the use of alternative resistor ratios. The following alternative circuits were built and the performance measured:

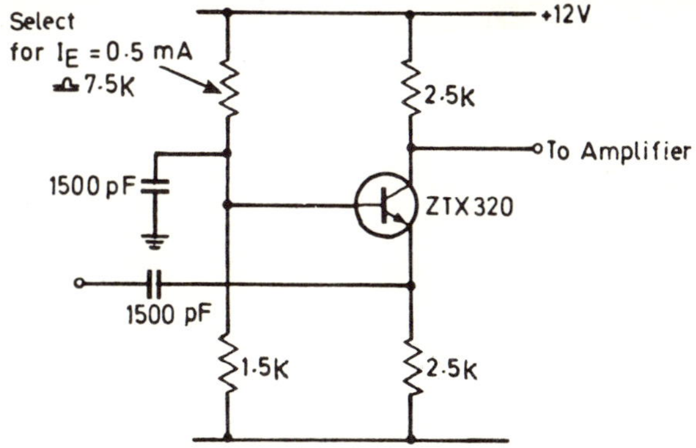

Fig. 2-7. Active input-matching stage.

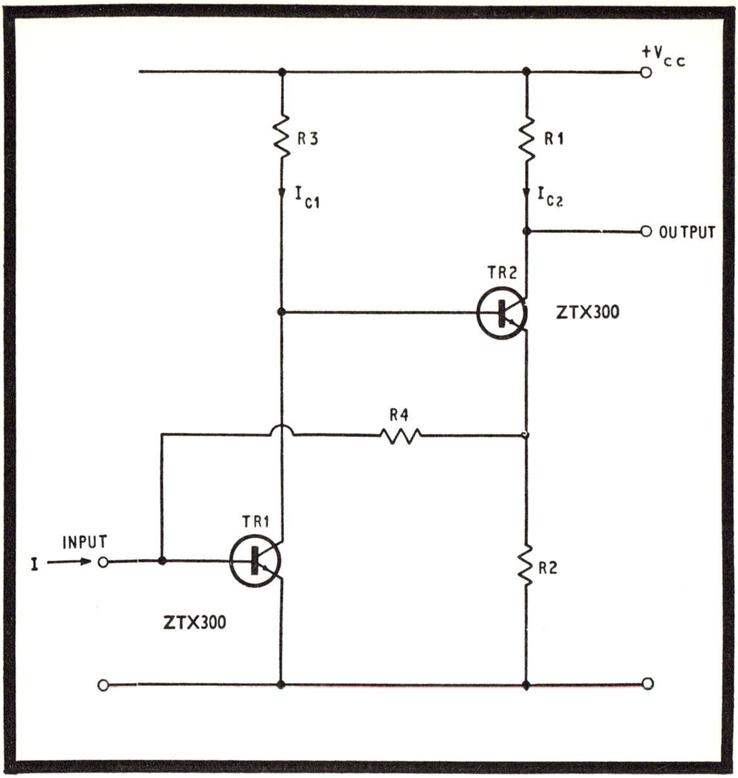

Fig. 2-8. Wideband current amplifier.

(a) R_F = 2.7K; R_S = 82 ohms
Current gain = 25
2 dB bandwidth = 130 MHz
Compensation capacitors: C_F = 0, CS = 11 pF.

(b) R_F = 4.7K, R_S = 82
Current gain = 42
3 dB bandwidth = 50 MHz
Compensation capacitors C_F = 0, CS = 22 p_F + (4.20 p_F trimmer).

As the frequency compensation required is dependent on circuit layout, the values of capacitors given are approximate. For optimum flatness of response, a sweep generator and visual display should be used and the capacitors adjusted accordingly.

The low input impedance is a major advantage when stages are used in cascade. The interstage capacitance has little effect on the overall bandwidth.

WIDEBAND CURRENT AMPLIFIER

The circuit in Fig. 2-8 has a low input impedance and a relatively high output impedance, with a bandwidth of over 6 MHz.

The design procedure is as follows:

1. Choose supply voltage, say, maximum 25V for ZTX300 or BCW10.

2. Choose Tr2 collector current. For maximum f_T, this should be between 3 mA and 20 mA. For maximum transfer resistance, V_{out} divided by I_{in}, I_{c2} should be as small as possible to allow R1 to be large. This will reduce bandwidth

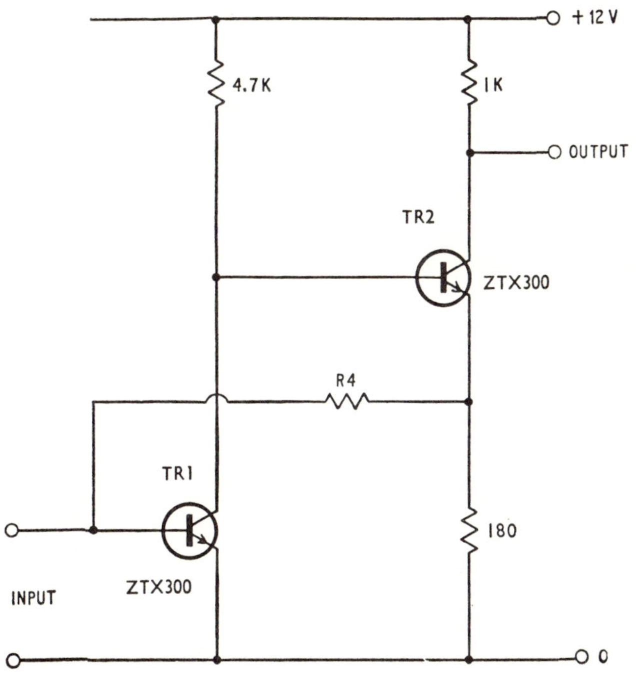

With R_4 = 1.8K ; bandwidth = 9 MHz, current gain = 10.
= 3.6K ; bandwidth = 6.5 MHz, current gain = 20.

Input impedance ≈ 15 ohms.

Output impedance = 1k.

Fig. 2-9. Final circuit for wideband current amplifier.

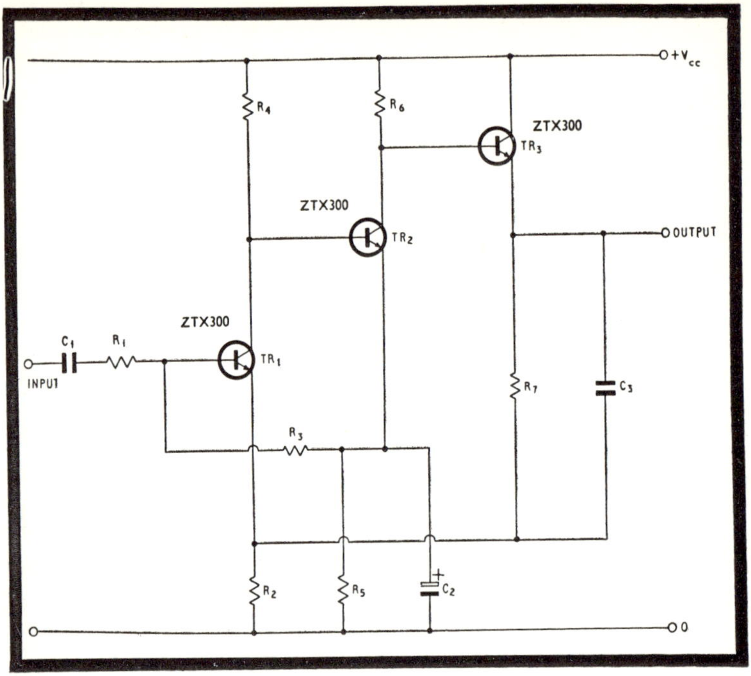

Fig. 2-10. Wideband voltage amplifier.

caused by stray capacitance and reduce f_T. Let's set I_{c2} at 5 mA.

3. Choose R1 to drop about 1V less than half of the supply volts (to allow for potential drop across R2). Therefore, R1 equals 1K.
4. The emitter-base junction of Tr1 will have a potential drop approximately 0.7V and R4 will drop a further 0.1 to 0.2V. Therefore, choose R2 to drop 0.9V; R2 will be 180 ohms.
5. Choose R3 to be about 5 times R1. Therefore, R2 equals 4.7K.
6. Choose R4 for current gain required.

Current gain AI R4/R2, or more accurately, (R4/R2) + 1.
Therefore, for current gain of 10, R4 ≈1.8K;
for current gain of 20, R4 ≈ 3.6K.

NOTE: Maximum value of R4 is that which will drop approximately 0.3V; i.e., 10K with minimum gain transistors in this case.

Final circuit is shown in Fig. 2-9.

WIDEBAND VOLTAGE AMPLIFIER

This will be an amplifier with a high input impedance, a low output impedance, and a bandwidth of 14 MHz. See Figs. 2-10 and 2-11.

1. Choose supply voltage, say 10V.
2. Choose Tr3 collector current. A value of 10 mA will be used as an example because the current gain of the transistor reaches a peak at this current.
3. Choose R7 + R2 to drop half the supply voltage; also, the ratio

$$\frac{R7 + R2}{R2}$$

gives the voltage gain. Thus, R7 + R2 500 ohms and, for a gain of approximately 20, R7 470 ohms, R2=22 ohms.
4. Choose R6 as 5 to 10 times R7. The higher value will give higher open-loop gain and lower distortion; the lower value will give a greater bandwidth. We'il use R7 times 10; that is, R6 = 4.7K.
5. Choose Tr2 collector current to drop half the supply voltage across R6; therefore, I_{c2} =·1 mA.
6. Choose R5 to drop approximately 1V. The potential drop (p.d.) across R2 is 0.25V in this case the V_{be} of Tr1 is estimated to be 0.6V, and the p.d. across R3 is estimated to be 0.1V. Adding, we have 0.25 + 0.6 + 0.1 equals 0.95 V. Therefore, R5 is 1K.
7. Choose R4 about 5 times R6; therefore, R4= 22K.
8. Calculate collector current of Tr1

$$= \frac{V_{cc} - 1.7}{R4} = 380 \text{ uA}$$

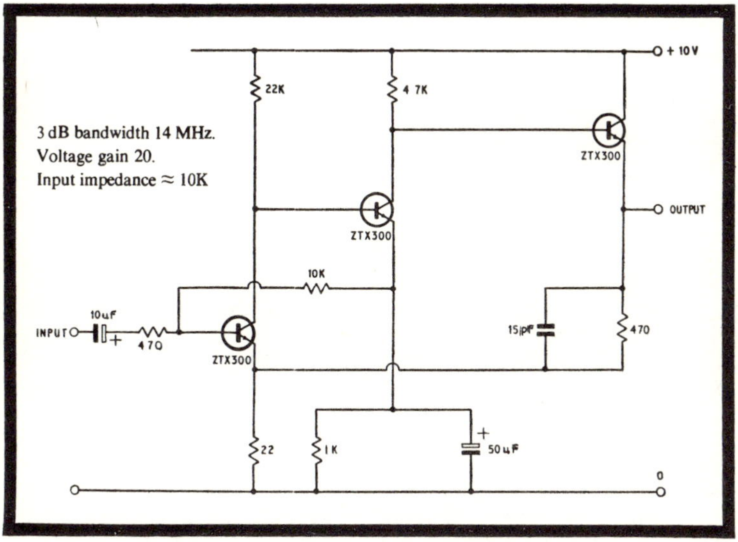

Fig. 2-11. Final circuit for wideband voltage amplifier.

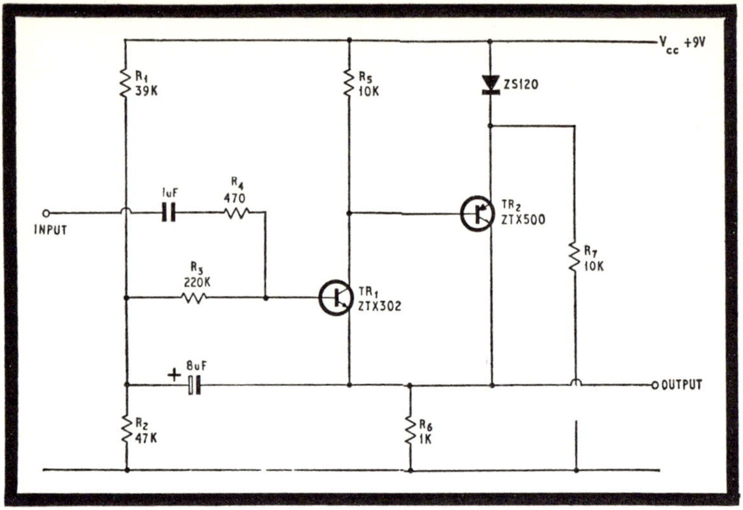

Fig. 2-12. High-impedance, unity-gain amplifier.

$$\text{Hence, the base current of Tr1} = \frac{380\ \text{uA}}{h_{FE}\ (\text{min})} = \frac{380}{30} = 13\ \text{uA}$$

9. Choose R3 to drop not more than 0.2V. If R3=10K, the p.d. caused by base current=0.13V, which is satisfactory.
10. Choose capacitor values. For wideband amplification, C2 is greater than 50 uF, C1 is greater than 10 uF. Capacitor C3 is necessary to remove a peak in the frequency response at high frequencies. In this example, a value of 15 pF was found to give maximum flatness.
11. Resistor R1 is a parasitic stopper. A value of 470 ohms was found to be adequate for the experimental circuit. With more careful construction, it may not be necessary.

HIGH IMPEDANCE, UNITY GAIN AMPLIFIER

In Fig. 2-12, the value of R6 is a compromise between input impedance and bandwidth. For the value shown (1K), the input impedance is greater than 1M and the bandwidth greater than 1 MHz at 1V rms output. By increasing R6, a higher input impedance can be obtained at the cost of reduced bandwidth.

Transistor Tr2 operates at a collector current of 4 mA; to produce this collector current, a base current of 35 uA is required (average). The bleed resistor, R5, takes 120 uA, so the total current drawn from Tr1 is 155 mA (average). Therefore, Tr1 base current is approximately 2 uA. This causes a drop or approx. 0.5V across R3 and the effective

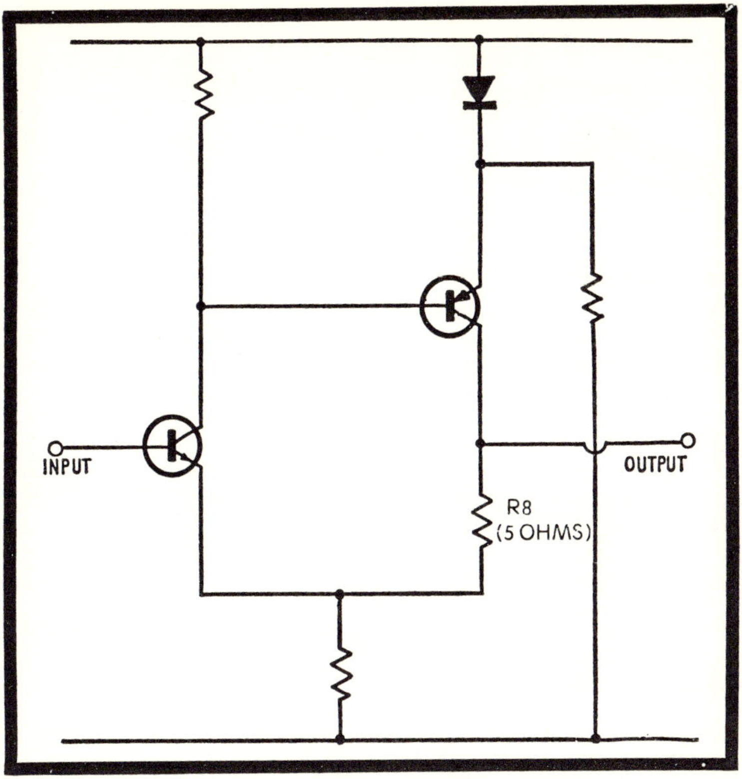

Fig. 2-13. Part of unity-gain amplifier.

resistance of the potential divider formed by R1 and R2. The values of R1 and R2 are not important, but the effective resistance of R1 and R2 in parallel should be large compared with R6 and small compared with R3. Resistors R1 and R2 are chosen to give a potential of 5V at the junction with R3. This gives approximately 4V across R6.

Performance measurements revealed that this circuit has an input impedance of at least 1M (depending on the current gain of the transistors used).

$$Z_{in} \approx (h_{fe1} \times h_{fe2} \times R6).$$

Bandwidth is greater than 1 MHz at 1V rms output. Notice that bandwidth here means the frequency at which waveform distortion is just visible, **not** the —3dB point, which is at a higher frequency.

Voltage gain of the test circuit was 0.995. This can be increased to exactly unity by the modification shown in Fig. 2-13. The value of R8 depends on the current gains of the transistors used.

Chapter 3

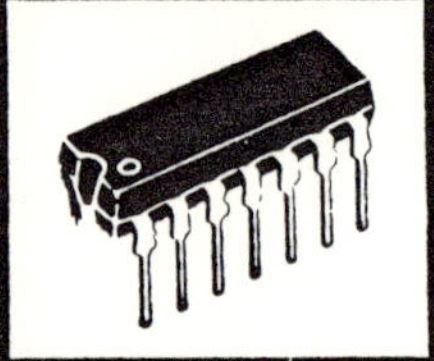

Astable Multivibrators

The astable multivibrator (Fig. 3-1) is basically an ac coupled two-stage amplifier with phase inversion over each stage and with the output coupled back to the input. This positive feedback causes the circuit to oscillator continuously between two quasi-stable states. Supose that Tr1 has just turned on and Tr2 is off. The voltage at the base of Tr2 rises exponentially toward the supply voltage as C2 charges through R2 until it reaches 0.7V, when Tr2 starts to conduct. The collector voltage of Tr2 starts to fall, and this change is coupled through C1 to the base of Tr1. Transistor Tr1 now begins to turn off and its collector voltage rises, increasing the base current of Tr2 and further reducing its collector voltage. This regenerative action results in very rapid switching from the "off" state to the "on" state and vice versa. The regenerative action ceases when Tr2 saturates. Transistor Tr1 is now completely cut off, with its base reverse biased by (approximately) - V_{cc}. The base potential of Tr1 now rises exponentially with a time constant of (C1 x R1) as C1 charges towards the supply voltage ($+V_{cc}$). As the base voltage at Tr1 reaches 1.7V, Tr1 switches on, Tr2 is cut off, and the cycle repeats.

CALCULATING ON-OFF TIMES

If the transistor leakage currents are ignored and the collector saturation voltage and base-emitter "on" voltage are assumed to be zero, simple approximate formulas for the periods t1 and t2 may be derived as follows:

The instantaneous base voltage of t1, V_{b1}=

$$2V_{cc}\left[1-\exp\frac{-t}{C1R1}\right]-V_{cc} \qquad = V_{cc}\left[1-2\exp\frac{-t}{C1R1}\right]$$

For $V_{b1}=0$ $\quad 2e^{-\frac{t}{C1R1}} = 1.$

$$t = t1 = C1R1 \log_n 2 \qquad (3\text{-}1)$$

i.e.: $\quad t1 = 0.693\ C1R$

$\quad t2 = 0.693\ C2R$

similarly,

t = 0.7CR is a useful approximation.

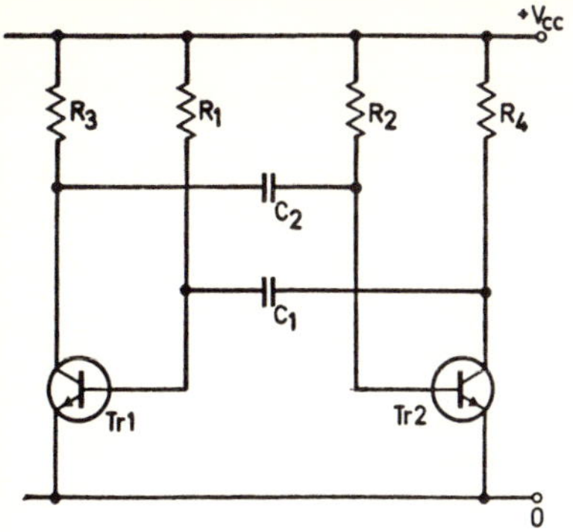

Typical component values:

R1 = R2 = 15K
R3 = R4 = 1K
C1 = C2 = 0.047uF
Tr1 = Tr2 = ZTX300
$+V_{cc} = 5$ v

Oscilloscope trace

(a)—Tr1 collector
(b)—Tr2 collector
(c)—Tr1 base
(d)—Tr2 base

Frequency = 963 Hz (measured)

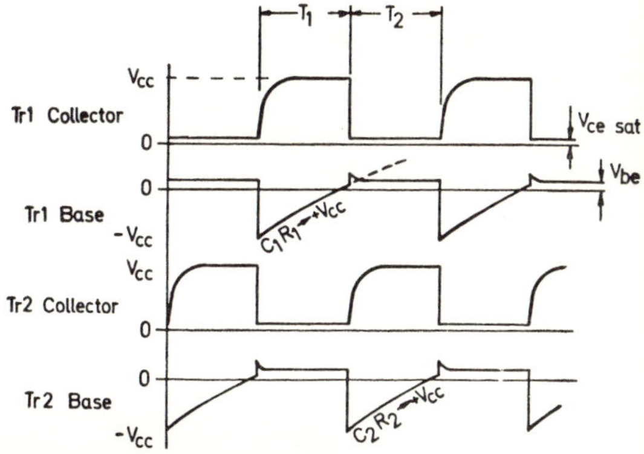

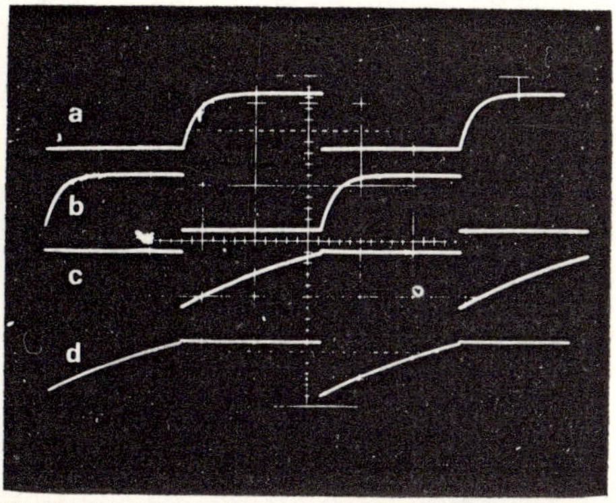

Fig. 3-1. Basic astable multivibrator.

These formulas are sufficiently accurate for most practical purposes; where a more precise estimate is required, the V_{cc}, V_{be}, and $V_{ce\,(sat)}$ must be taken into account. Periods t1 and t2 can be expressed more accurately as:

$$t1 = C1\ R1\ \log_n \frac{2V_{cc} - V_{be} - V_{ce\,(sat)}}{V_{cc} - V_{be}} \qquad (3\text{-}2)$$

and similarly for t2.

For the circuit shown in Fig. 3-1, the frequency calculated from Eq. 3-1 is 1.022 kHz. Assuming $V_{be} = 0.7V$ and $V_{cc\,(sat)}$ 0.1V, Eq. 3-2 gives — $t1 = t2 = C1\ R1\ \log_n \frac{10-0{\cdot}7-0{\cdot}1}{5-0\ 7}$

$$= C1R1 \log_n 2.14$$
$$= 0.047 \cdot 10^{-6} \cdot 15.10^3 \cdot 0.761 \text{ sec}$$
$$t1 = t2 = 0.536 \text{ msec}$$

Hence, the total period is 1.072 msec and the frequency is 934 Hz. The measured frequency of the actual circuit was 963 Hz with 1 percent components.

RESISTOR VALUES

The maximum values of R1 and R2 are limited to h_{fe} x R3, and h_{fe} x R4, respectively, where h_{fe} is the minimum direct-current gain of the transistors used. In practice, it is advisable to use a value somewhat lower than this to ensure that the transistor is fully saturated.

The minimum values of R1 and R2 must be such as to allow the collector voltage to complete its full excursion during the pulse period. Referring to Fig. 3-1, for Tr1 collector to recover to at least $0.98V_{cc}$, t1 must be at least equal to 4C2R3 since $(1\text{-}e^{-4}) = 0.98$.

Hence, $t1 = 0{\cdot}69\ C1R1 \geq 4{\cdot}C2R3$,
if $C1 = C2$, $R1 \geq 5.8R3$. (3-3)
and $R2 \geq 5.8R4$.

If recovery to 0.99 V_{cc} is required, these become:

$R1 \geq 6.7R3$
$R2 \geq 6.7R4$ (3-4)

CIRCUIT MODIFICATIONS

As described so far, the basic astable multivibrator has two disadvantages which can, where necessary, be overcome by modifications to the circuit.

Sharpening the Response

The positive-going edge of the collector waveform is slow because the timing capacitor has to recharge through the collector load resistor. This can be avoided by providing a separate recharge path for the timing capacitor, as shown in Fig. 3-2. When Tr2 turns off, the diode is reverse biased and the

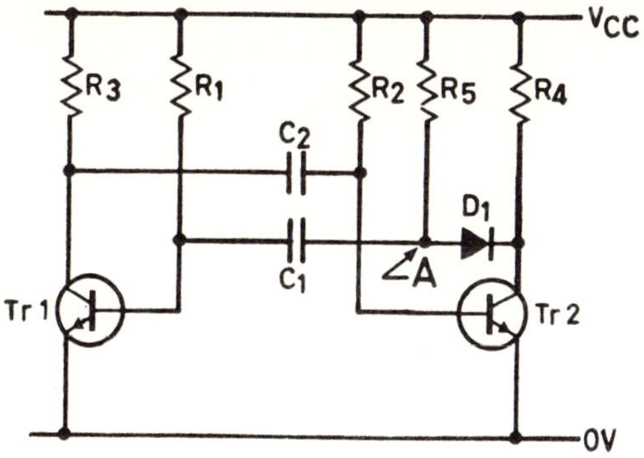

Typical component values:

R1	=	R2	=	15K
R3	=	1K		
R4	=	R5	=	2200
C1	=	C2	=	0.047uF
Tr1	=	Tr2	=	ZTX300
D1	=	ZS140		
$+V_{cc}$	=	5 volts		

Trace (a) Tr1 collector
(b) Tr2 collector
(c) Point "A"
(d) Tr2 base

Frequency = 1.066 kHz (measured)

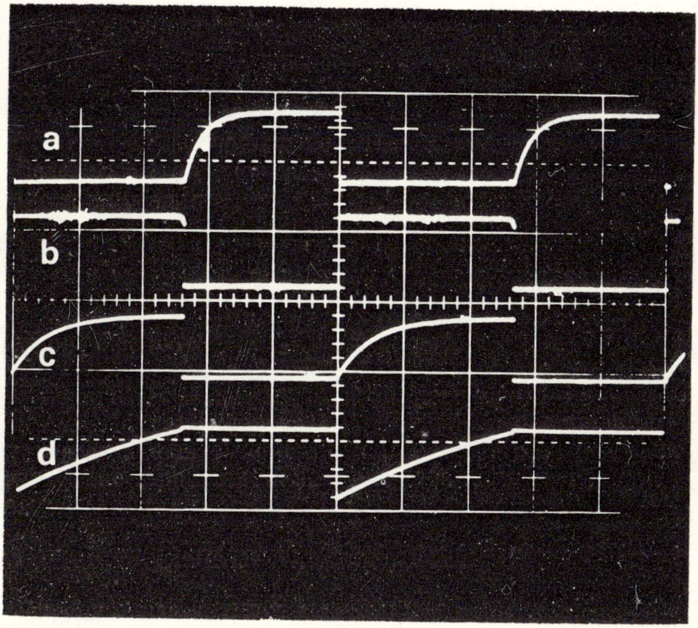

Fig. 3-2. Astable multivibrator modified for improved waveform.

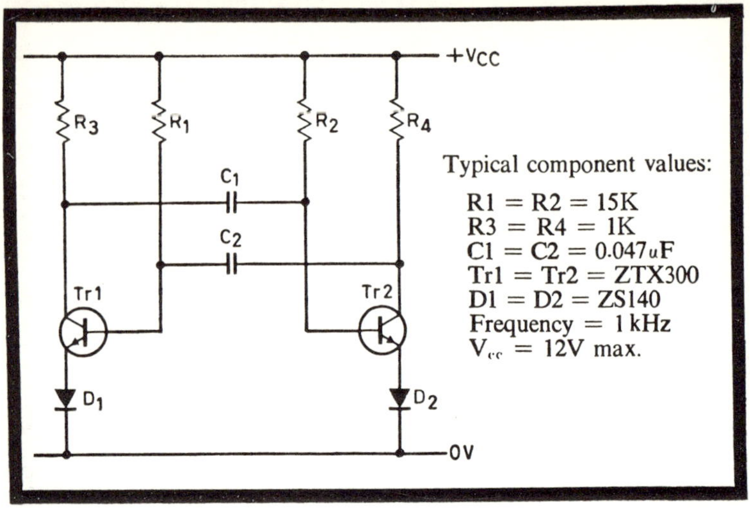

Fig. 3-3. Astable multivibrator protected against base-emitter breakdown.

collector voltage rises rapidly to V_{CC} while the voltage at point A rises exponentially with time constant C1 x R5 as C1 recharges. When Tr2 turns on, the diode becomes forward biased and R4, R5 appear effectively in parallel. It is usually convenient for the values of R4 and R5 to be equal, but this is not important as long as R5 is small enough to completely recharge C1 during the period that Tr2 is turned off. The technique, of course, can be applied to both sides of the multivibrator when clean waves are required at both outputs.

Voltage Limitations

In the basic multivibrator of Fig. 3-1, the transistor base is momentarily reverse biased to $-V_{CC}$. This limits the maximum value of V_{CC} to the base-emitter breakdown voltage of the transistor (usually 5V). This is too low for many applications; however, higher supply voltages can be used if the transistor is protected against base-emitter breakdown. Two circuits which do this are shown in Figs. 3-3 and 3-4.

In Fig. 3-3, either diode (D1 or D2) becomes reverse biased when the base of its transistor is driven negative, and this prevents any current from flowing in the base-emitter junction. The diodes should preferably be high-speed types, such as ZS140, and the transistors chosen must have a collector-base breakdown voltage (V_{cbo}) of at least twice V_{CC}, the collector being at $+V_{CC}$ when the base is reverse biased to minus V_{CC}.

The maximum value of V_{CC} in the circuit of Fig. 3-4 is limited to the collector-emitter breakdown voltage (V_{ceo1} of the transistor because D3 and D4 completely isolate the transistor bases. In this circuit, it is often advantageous for D1 and D2 to be fast types (e.g., ZS140) and D3 and D4 to be slow types (e.g., ZS120) as the stored charge in D3, D4 helps to turn the transistors off rapidly. At low speeds, D1 and D2 can be omitted if fast diodes are used for D3, D4.

Another technique which is particularly useful at high repetition rates is to "catch" the collector voltage at 5V by means of D1 and D2 and zener D3, as shown in Fig. 3-5. In this case, the collector voltage swing, and hence the reverse bias applied to the base, is limited to 5V and the collector waveform is considerably improved. Note, however, that the simplified formulas derived earlier no longer apply, and the component values required must be calculated from first principles.

All of these circuits can, of course, be used with pnp transistors if all diodes and the supply voltage are reversed. All of the waveforms produced will also be inverted. Many of the techniques described here can also be applied to monostable multivibrators.

The types of astable multivibrator described so far are the most generally used, but other types, described below, are useful in some circumstances.

THE CURRENT-MODE ASTABLE

In this circuit, the transistors do not saturate. This greatly reduces the stored charge and enables the circuit to operate at

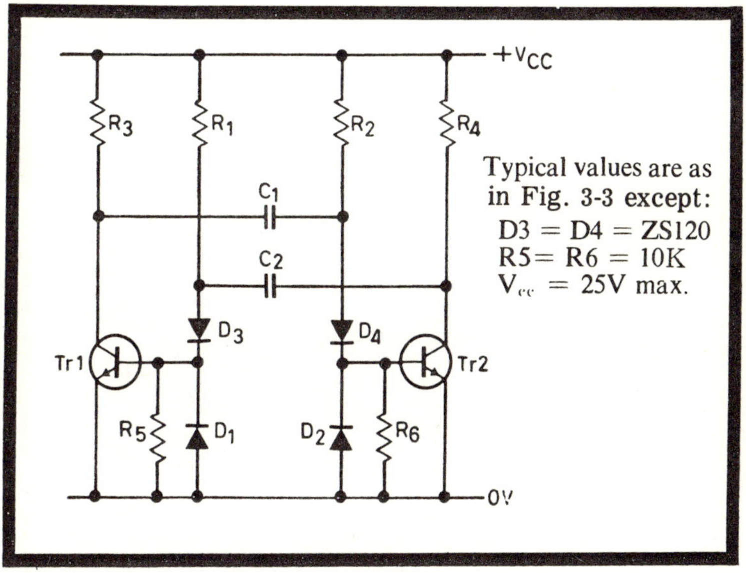

Fig. 3-4. Astable multivibrator protected against base-emitter breakdown by diodes that isolate the bases.

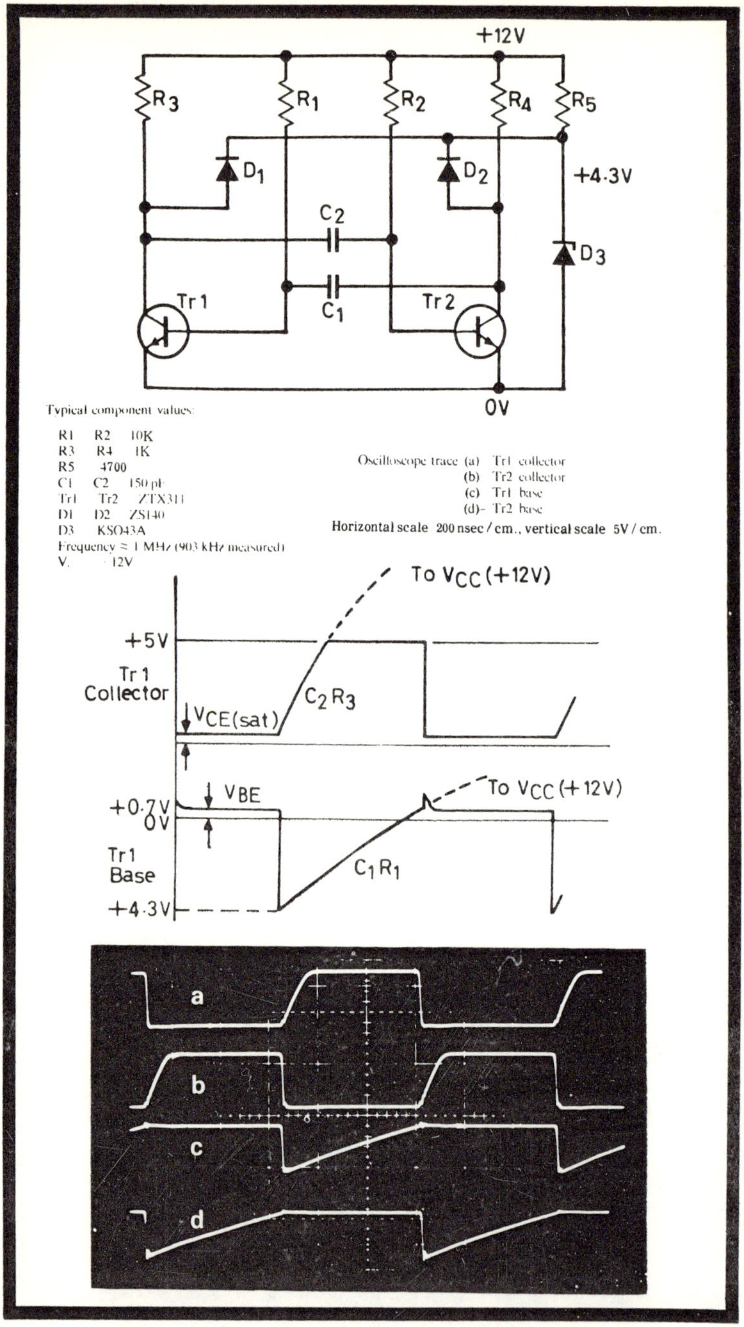

Fig. 3-5. Astable multivibrator with catching diodes.

much higher repetition rates than are possible with the simple saturating astable. The basic circuit is shown in Fig. 3-6.

Assume that Tr1 has just been turned off and that Tr2 is on. The potential at Tr1 emitter falls linearly as C charges with I until the emitter reaches −0.7V, when Tr1 starts to conduct. The collector voltage of Tr1 starts to fall and Tr2 is reverse biased. Both current sources, I1 and I2, now flow into the emitter of Tr1, causing a voltage drop of (I1 + I2) x R across the collector load resistor. This voltage drop must not be sufficient to saturate Tr1. The voltage at Tr2 emitter now falls linearly as C discharges with I2 until it reaches 0.7V

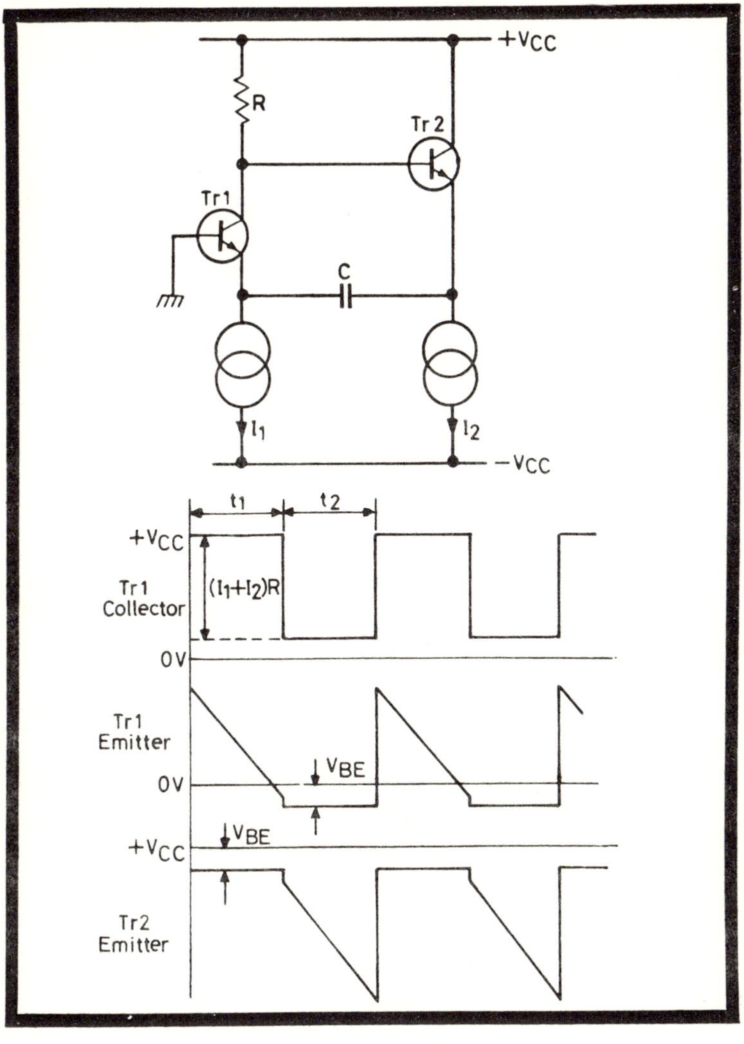

Fig. 3-6. Basic Bowes-type astable multivibrator.

below the collector voltage of Tr1. Transistor Tr2 then starts to conduct, the current flowing into Tr1 emitter is reduced, and the collector voltage of Tr1 rises until Tr1 is completely cut off and Tr2 is conducting I1 and I2. The cycle then repeats. Timing equations can be derived as follows, assuming that the transistor common-base current gain is unity and that the leakage currents and V_{be} can be neglected:

$$\text{Voltage drop } V1 = (I1 + I2) \times R = \frac{I1 \times t1}{C} = \frac{I2 \times t2}{C}$$

(since CV=IT)

$$\text{Therefore, } t1 = \frac{(I1 + I2)\ CR}{I1}$$

$$\text{and } t2 = \frac{(I1 + I2)\ CR}{I2}$$

$$\text{Hence, the total period } t1 + t2 = \frac{(I1 + I2)^2\ CR}{I1\ \ I2}$$

If $I1 = I2$, then $t1 + t2 = 4CR$;

or if $\frac{I1}{I2} = n$, for unequal currents, $t1 + t2 = \frac{(1 + n)^2\ CR.}{n}$

A practical circuit is shown in Fig. 3-7, together with the waveform obtained. Figure 3-8 shows the effect of replacing the current sources with resistors delivering the same average currents. As can be seen, there is some degradation of the waveform and the frequency is increased because of the exponential charging.

SERIAL ASTABLE CIRCUITS

Serial-type astables, in which the two transistors are connected in series, have been described by various authorities. Usually the design has been aimed at obtaining a better output waveform than that provided by the conventional astable.

Figure 3-9 shows a slightly modified version of the Smith astable. With ZTX300 transistors in series, the circuit will oscillate up to 100 kHz, and to over 1 MHz with ZTX310 transistors in series. The transition times are approximately 50 msec on both rise and fall edges. Output voltage swing with a 15V supply is approximately 10V p-p, the lower level being 2.5V and the upper level 12.5V.

Frequency of oscillation with the component values shown is 3 divided by CR. The timing resistors are R1 and R2.

Figure 3-10 shows a series Ristic-type astable. Its performance is similar to the Smith circuit, except that the output voltage swing is reduced to about 5V with a 15V supply. Rise and fall times are approximately 100 msec, and the frequency of oscillation is given by

$$f_0 = \frac{1.35 \times 10^{-4}}{C}$$

with the component values shown.

Figure 3-11 shows a Ho-type complementary series astable. Very large mark-space ratios are possible with this circuit, which has two identical but complementary outputs. The pulse period is not well defined because it depends on the transistor h_{fe}, but the period between the pulses is given by the normal astable formula; that is, t equals 0.7CR. It is important that the base current provided by R1, R2 should **not** be sufficient to saturate the transistors, or latch-up will occur.

FOUR EXAMPLES FOR WORST-CASE DESIGN

The following factors should be remembered when performing a worst-case design:

(a) The V_{be} of a silicon transistor falls by approximately 2 mV

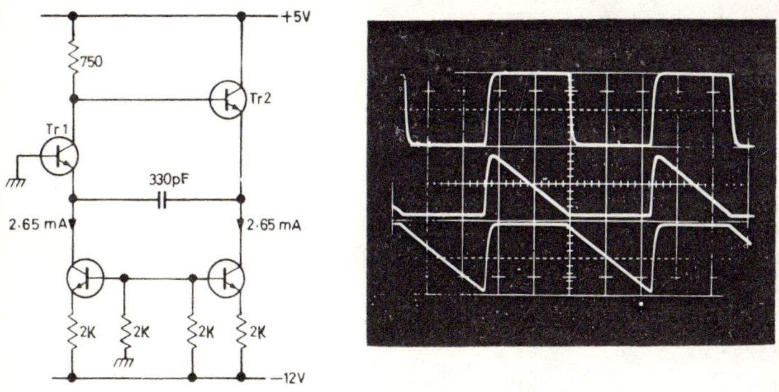

Practical circuit for 1 MHz
All transistors—ZTX311
Measured frequency—1.09 MHz

Upper trace, Tr1 collector
Center trace, Tr1 emitter
Lower trace, Tr2 emitter
Vertical scale: 2V/cm
Horizontal scale: 200 nsec/cm

Fig. 3-7. Practical Bowes-type astable multivibrator.

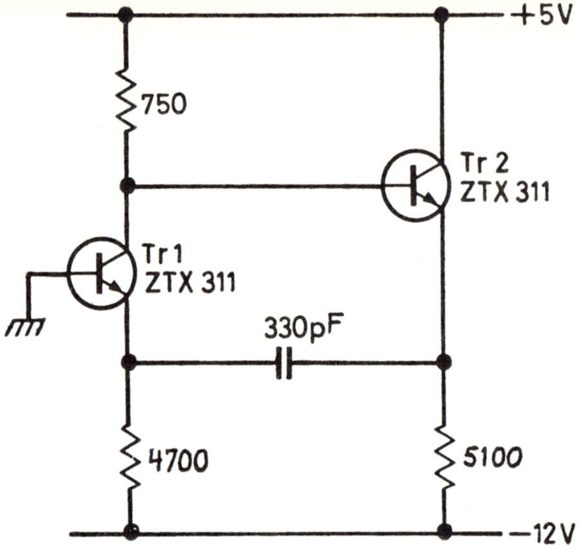

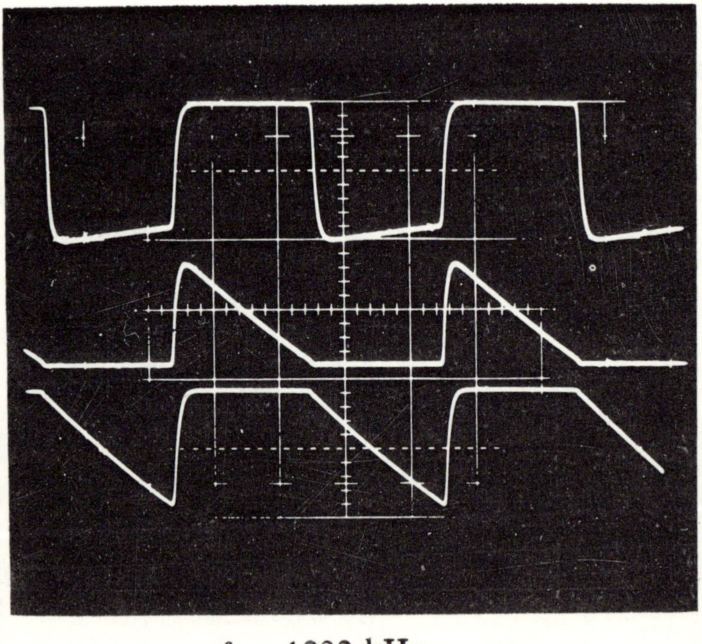

f = 1232 kHz

Fig. 3-8. Bowes-type astable multivibrator with resistor current sources.

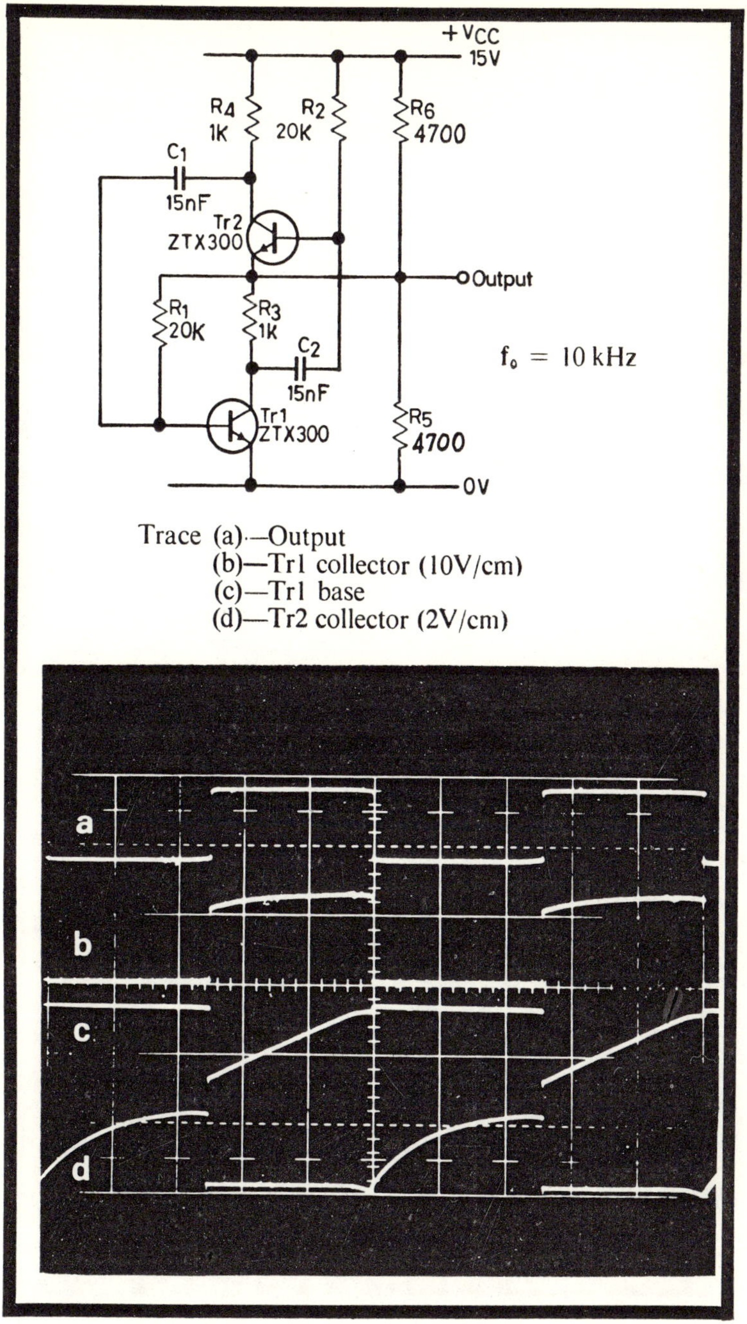

Fig. 3-9. Modified Smith-type astable multivibrator, with waveforms.

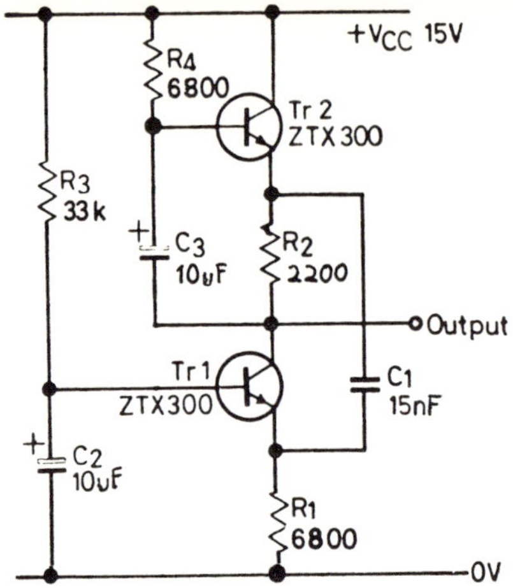

C2 and C3 are decoupling capacitors.

$f_o = 9$ kHz

Upper trace—output
Center trace—Tr1 emitter
Lower trace—Tr2 emitter
Vertical scale = 5V/cm
Horizontal scale = 20 usec/cm

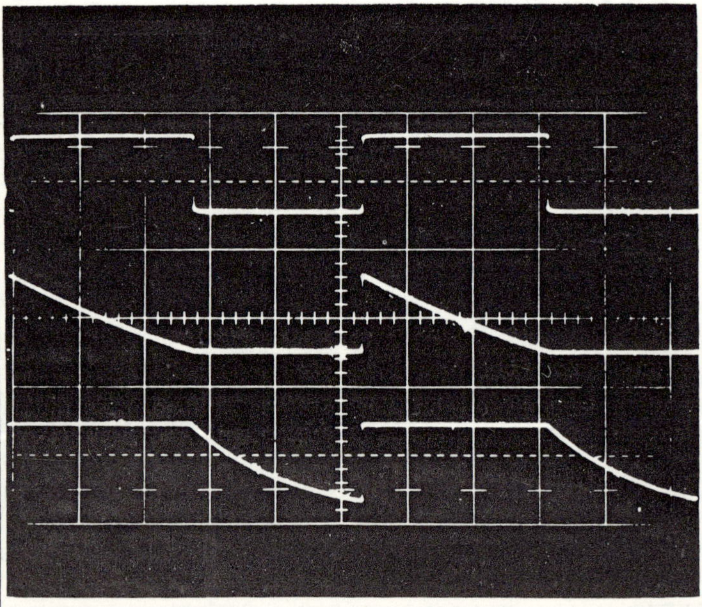

Fig. 3-10. Ristic-type astable, multivibrator with waveforms.

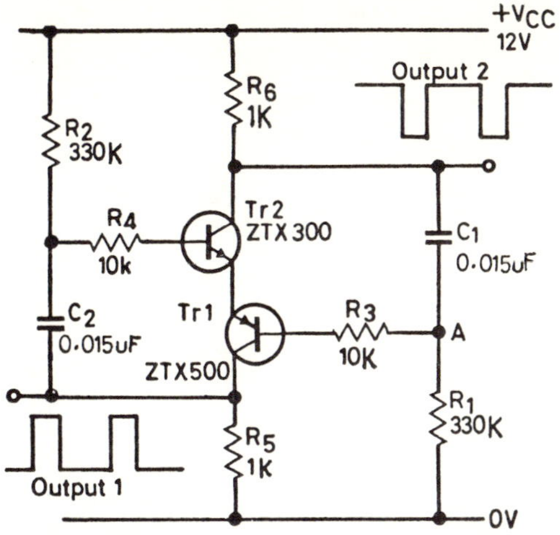

Upper trace—Output 2
Centre trace—Output 1
Lower trace—Point A

Vertical scale—5V/cm
Horizontal scale—1 msec/cm

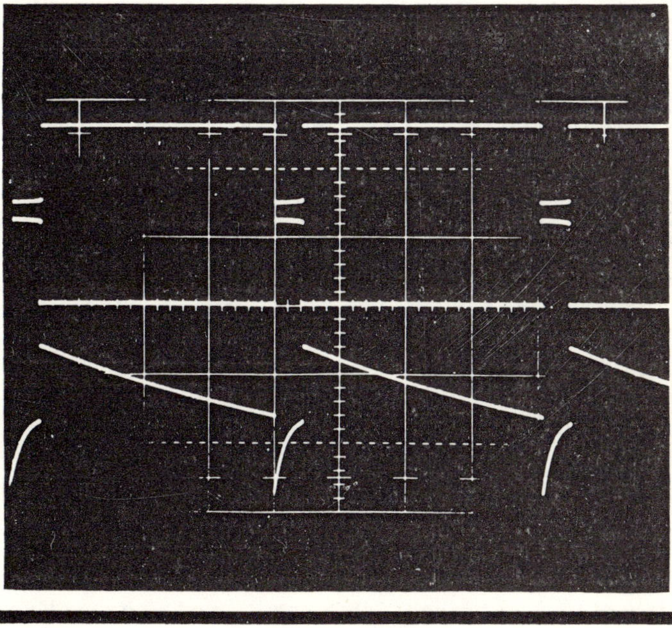

Fig. 3-11. Ho-type astable multivibrator, with waveforms.

for every 1 deg C increase in temperature.
(b) The h_{fe} of a silicon transistor rises by approximately 0.7 percent for every 1 deg C increase in temperature.
(c) Depending on the conditions of use, the value of a resistor will change with time because of the effects of soldering, moisture, and aging, and may go outside its nominal tolerance. A molded carbon resistor may change by 20 percent from its original value, a carbon-film resistor by 5 percent, and a metal-oxide resistor by 1 percent. In the examples given, it will be assumed that a 10-percent resistor will have an ultimate tolerance of 20 percent, etc.

Example 1

A 1 kHz square wave astable to operate from a 5V plus or minus 0.5V supply is to be designed. No dc load, but output impedance of 1K. Ambient temperature —10 to +50 deg C. (Circuit and waveforms are shown in Fig. 3-1.)
1. Choose transistor type ZTX300.
2. Choose collector load resistor, 1K nominal, 20 percent.
3. Minimum supply voltage is the worst condition for turning the transistor fully on. With a 4.5V dc supply, collector current equals 4.5 mA nominal and equals 5.6 mA maximum with R3, R4 20 percent low.
4. Minimum current gain of ZTX300 at 5.6 mA is 45 at 25 deg C and is 34 at —10 deg C.
Hence, the minimum base current to saturate equals $\frac{5.6}{34}$, equals 0.165 mA with base resistor 20 percent high.
5. From ZTX300 curves, V_{be} at —10 deg C equals 0.8V.
Hence, the maximum value of base resistor R1, R2 =

$$\frac{4.5-0.8}{0.165} \times \frac{100}{120}$$; that is, the

maximum value of R1, R2 equals 18.7K.
6. The minimum value of R1, R2 to ensure recovery of the collector voltage to $0.99V_{cc}$ equals 6.7K (from Eq. 3-4). Any value of timing resistor R1, R2 between these two limits will be suitable. For a 1 kHz square wave output, both periods T1 and T2 must be 0.5 msec.

$$t=0.7CR;\text{ therefore, }CR=\frac{0.5}{0.7}=0.7\text{ msec.}$$

The above conditions are satisfied if R1=R2=15K and C1=C2=0.047 uF.

Example 2

Design an astable to generate a 100 usec pulse with a repetition time of 1 msec. Other conditions are as for Example 1.

1. Choose transistor type ZTX300.
2. Choose load resistor R4 equals 1K + 20 percent.
3, 4, 5. Maximum value of R2 equals 18.7K (as in Example 1).
6. Transistor Tr2 is held off for 900 usec. Therefore, time constant C2R2 equals 900/0.7 or 1280 msec.
A reasonable approximation to this is obtained if R2 equals 18K, C2 equals 0.068 uF.
7. Resistor R3 must be able to fully charge C2 during the 100 usec pulse. From Eq. 3-3, for recovery to 98 percent of the supply voltage, R1C1 is greater than 5.8 times R3C2, but R1C1 = R2C2/9 because a 100 usec pulse is required from Tr1.
Therefore, R3 is less than

$$\frac{R2}{5.8 \times 9} = \frac{08K}{52} = 350 \text{ ohms}$$

The nearest preferred value below this is 330 ohms.
8. Worst-case collector current of Tr1, with a 4.5V supply, is

$$\frac{4.5}{330} \times \frac{100}{80} = 18.7 \text{ mA}$$

The minimum current gain of ZTX300 at 18.7 mA equals 54 at 25 deg C and is 40 at —10 deg C; therefore, minimum base current to saturate=

$$\frac{18.7}{40} = 0.47 \text{ mA (with R1 20 percent high}$$

9. Maximum value of R1=

$$\frac{4.5 - 0.8}{0.47} \times \frac{100}{120} = 6.6\text{K}$$

10. Minimum value of R1, R2C2 is greater than 5.8 x R4C1; 9 x R1C1 is greater than 5.8 x R4 C1; therefore, R1 is greater than

$$\frac{5.8R4}{9} = 650 \text{ ohms}$$

However, a value as low as this would be undesirable because it would impose an additional load on Tr2.

For a 100 usec pulse, $C1R1 = \frac{100}{0.7} = 140$ usec.

A reasonable approximation to this is obtained with R1 equal to 6200 ohms and C1 equal to 0.022 uF.

The finished circuit and output waveforms are shown in Fig. 3-12.

Example 3

Design an astable to switch on a 12V, 160 ohms relay for 3 sec and off for 3 sec. Supply voltage 12V, plus or minus 1V.

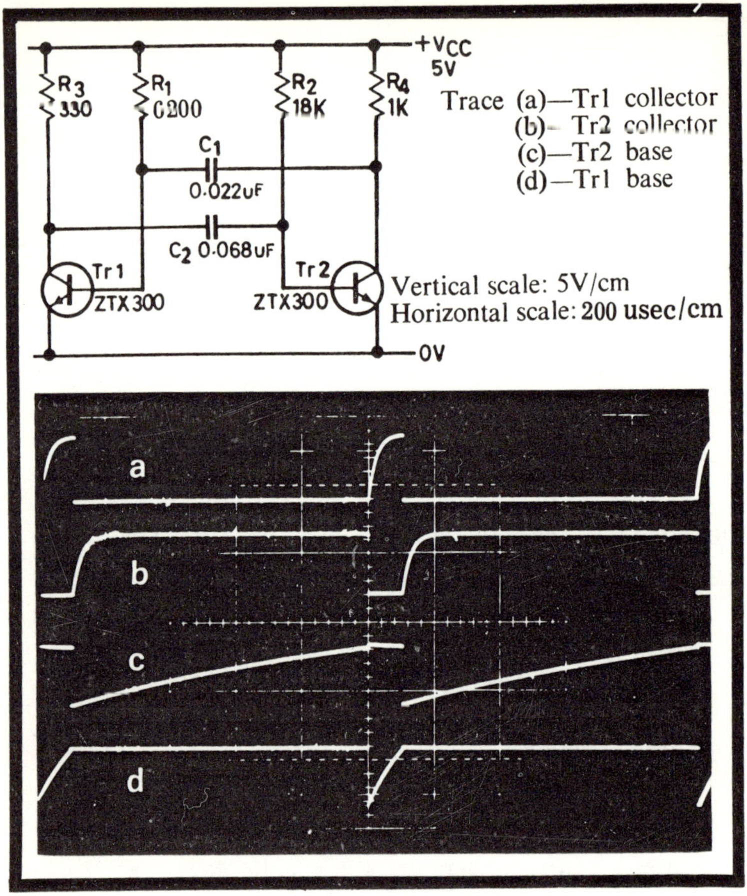

Fig. 3-12. The 1 kHz astable multivibrator.

Ambient temperature 0 to 50 deg C. Assume initially that a circuit similar to Fig. 3-12 will be used with the relay coil replacing R4.

1. With a 12V supply, protection diodes are required in the emitter of Tr1 and Tr2.
2. Choose transistor type ZTX300, which can switch the required voltage and current.
3. With 11V supply, assuming 20 percent tolerance on relay coil resistance:

maximum collector current of Tr2=

$$\frac{(11 - V_d - V_{ce\,(sat)}}{160} \times \frac{100}{80}$$

where

$$V_d = \text{diode voltage} = \frac{10}{160} \times \frac{100}{80} = 78\,\text{mA}$$

Minimum current gain of ZTX300 at 78 mA is 50 at 25 deg C and is 41 at 0 deg C.
Minimum base current required to saturate: 78/41 equals 1.9 mA, with R2 20 percent high.
4. Maximum value of R2: ′

$$\frac{(11 - V_d - V_{be}}{1.9} \times \frac{100}{120} = 4.1K$$

5. For Tr2 to be turned off for 2 sec, 2.86 sec.

If R2=3.9K, then C2=730 uF. This value is not impractical, but is inconveniently high. A lower value can be used if Tr2 is replaced by a Darlington pair, as shown in Fig. 3-13.
6. Referring to Step 3 above, Tr3 emitter current = 1.9 mA. Minimum current gain of ZTX300 at 1.9 mA is 35 at 25 deg and is 29 at 0 deg C.
Minimum base current required to saturate Tr3 with R2 20 percent high 65 uA.

7. Maximum value of R2: 110K.

If a value of 91K is chosen for R2, then a close approximation to the required time constant is obtained if C2=32 uF. Resistor R5 is chosen merely to limit the collector current of Tr3 to a safe value in excess of 1.9 mA. The R5 value of 2200 is suitable.

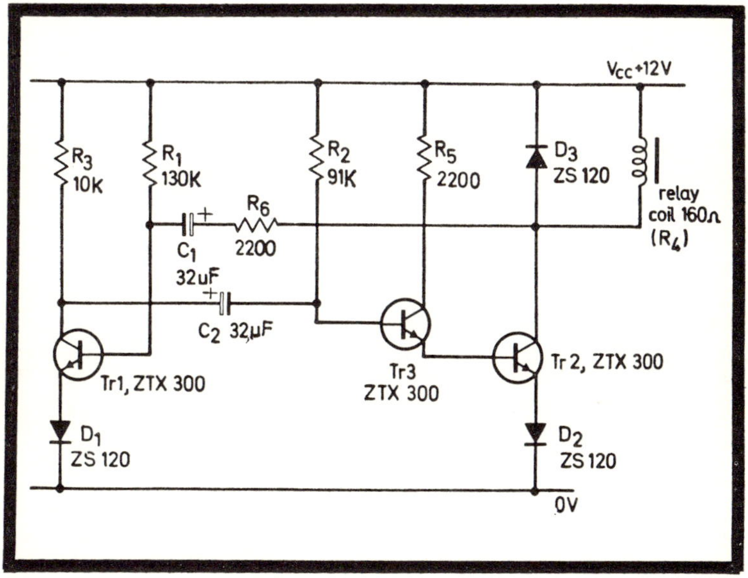

Fig. 3-13. A low-speed astable multivibrator to switch a relay on for 3 seconds and off for 2 seconds.

8. From Eq. 3-3, we see that R1C1 is greater than 5.8 times R3C2, but R1C1 equals 3/2 times R2C2 since Tr1 is off for 3 sec.
Hence, R2C2 is greater than 3.9 times R3C2. That is, R3 is less than R2/3.9 or 91K/3.9, which equals 23.3K.
This is the **maximum** value that may be used. In this design we will use a 10K for R3.

9. Maximum collector current of Tr1

$$= \frac{11 - V_d - V_{ce}\,(sat)}{10k} \times \frac{100}{80}$$

$$= 1.25 \text{ mA}$$

Minimum current gain of ZTX300 at 1.25 mA=27 at 0 deg C.
Minimum base current to saturate Tr1 with R1 20 percent high

$$= \frac{1.25}{27} = 46.3 \text{ uA}$$

10. Maximum value of R1

$$= \frac{11 - V_d - V_{be1}}{46.3} \times \frac{100}{120} \times 10^6$$

$$= 170K.$$

11. If R1 is made 130K a close approximation to the required time constant of 3/0.7 sec is obtained with C1 being a 32 uF.

The finished circuit is shown in Fig. 3-13. It was found necessary to include R6 in order to prevent spurious operation caused by inductive ringing in the relay coil, but it has virtually no effect on the timing. Diode D3 protects Tr2 from the inductive spike produced when Tr2 turns off.

Example 4

Design a wide-range, variable-speed astable for generating repetitive pulses. See Fig. 3-14.

No design procedure will be given for this circuit, as it is basically the same as Example 2; however, the design is complicated by the fact that the duration of one of the quasi-stable states must be variable over a wide range. A high-gain transistor is essential for Tr2, and type ZTX107 was chosen.

With the component values shown, t1 (= 0.7C1 x R1) is constant at 0.5 msec, but T2, defined by (R2 + R5), is variable over a range of 0.5 to 10.5 msec. This gives a range of repetition time of 1 msec to 11 msec. Other time ranges can be accommodated by the choice of suitable capacitor values.

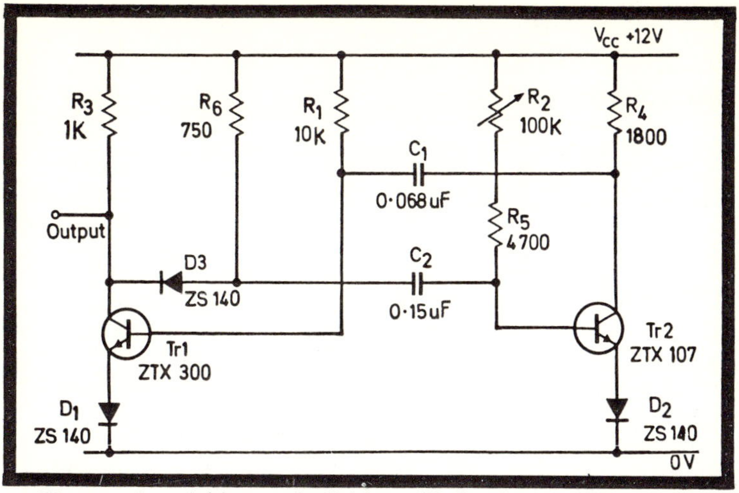

Fig. 3-14. A variable-speed astable multivibrator with a 10:1 range.

C1	C2	PRF
680 pF	1500 pF	10 usec to 110 usec
6800 pF	0.015 uF	100 usec to 1.1 msec
0.068 uF	0.15 uF	1 to 11 msec
0.68 uF	1.5 uF	10 to 110 msec

The output waveform at Tr1 collector is a rectangular pulse with rise and fall times of less than 50 usec.

Chapter 4

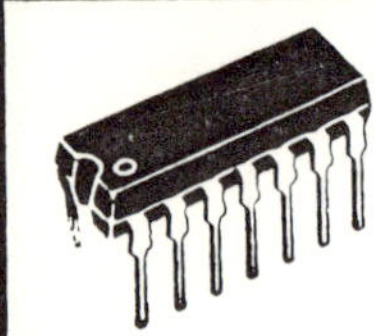

Monostable Multivibrators

A monostable multivibrator (Fig. 4-1) has one ac coupling and one dc coupling. This gives it one stable state in which it normally rests and a quasi-stable state into which it can be triggered by means of an external pulse.

The timing equations for the circuit shown in Fig. 4-1 are similar to those for the basic astable, the duration of the quasi-stable state being given by $t = 0.7\ C1R$. Capacitor C2 is a speedup capacitor which reduces the turnoff time of Tr2 by rapidly removing the stored charge. It does not normally influence the timing.

This circuit is triggered from the negative edge of the input waveform, and C3 must be as small as possible so that it does not influence the timing. Generally, C3 is less then one-tenth the value of C1.

In some cases, when the positive edge of the input waveform occurs before the end of the quasi-stable state, the circuit may reset prematurely. This can be prevented by the addition of a diode and a resistor at the base of Tr1, as shown in Fig. 4-2.

Figure 4-3 is a modification of Fig. 4-2 to enable the circuit to trigger on the positive edge of the input waveform. To ensure that the trigger capacitor recharges fully between trigger pulses, the time constant C3R5 should be much less than the duration of the positive part of the trigger waveform for Fig. 4-2 and the negative part of the trigger waveform for Fig. 4-3. It is usually convenient to make R5 about the same value as R1 or R3, although higher values can be used in low-speed circuits. For high-speed operation, a diode can be connected across R5, as shown in Fig. 4-3.

An alternative of triggering from the positive edge of the input waveform is shown in Fig. 4-4. In this circuit, the values of R5 and C3 should be so chosen that the duration of the trigger pulse at Tr3 collector is much shorter than the required pulse.

At low speeds, the values of C2 and C3 can be 10 to 1000 times less than C1. At high speeds, the minimum values should be determined experimentally using high-gain transistors. The graphs in Chapter 1 give the approximate values required, and these should be regarded as minimum.

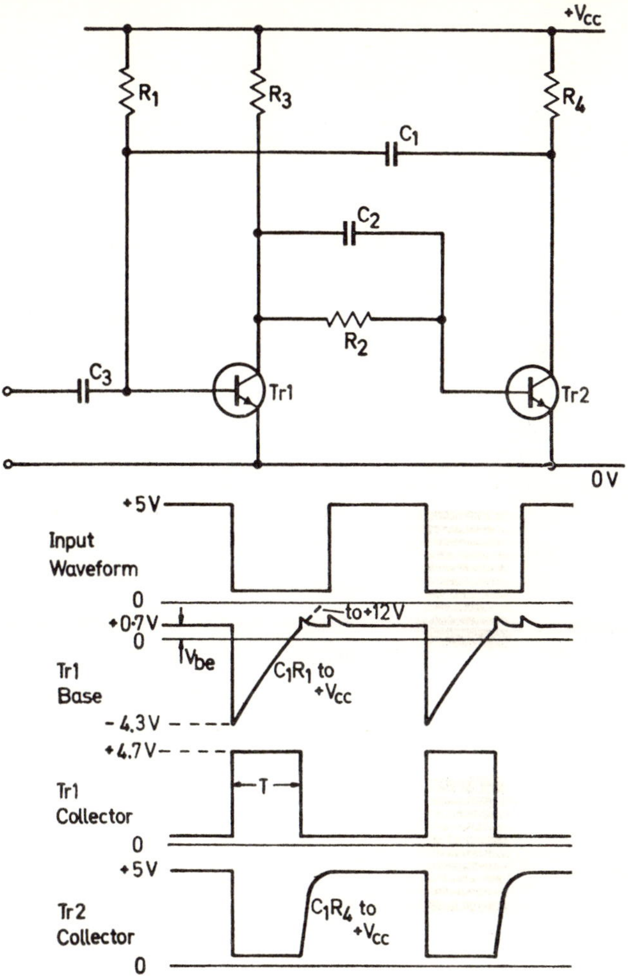

Typical component values:

$R1 = R2 = 15K$

$R3 = R4 = 1K$

$C1 = 0.1\mu F$

$C2 = 100\,pF$

$C3 = 150\,pF$ (minimum)

$Tr1 = Tr2 =$ ZTX300

Pulse width, $T = 1$ msec, $V_{CC} = +5V$.

Fig. 4-1. Basic monostable multivibrator, with waveforms.

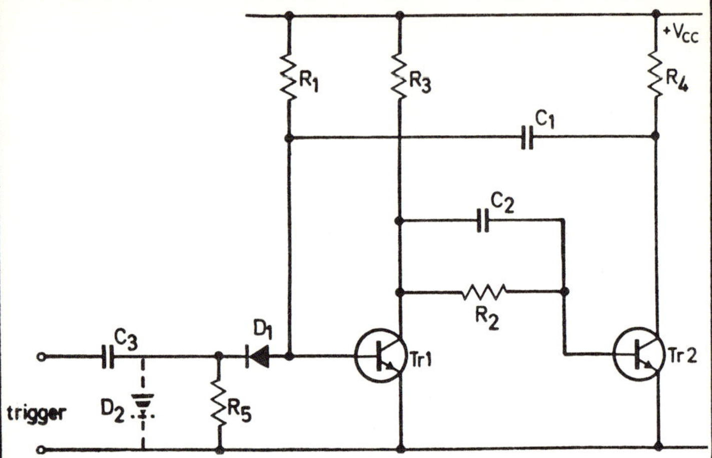

Typical component values as Fig. 4-1 except:

R5 = 15K
D1 = ZS140
D2 = ZS140 (if required)

Fig. 4-2. Negative-trigger monostable multivibrator.

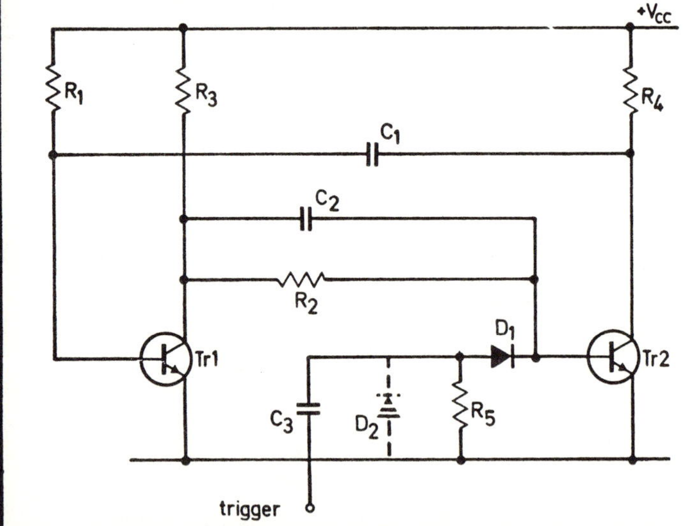

Fig. 4-3. Positive-trigger monostable multivibrator. Values and part numbers same as in Fig. 4-2.

With supply voltages greater than 5V, protection diodes are necessary for Tr1, as discussed in Chapter 3. Also, the output waveform from Tr2 has a slow positive edge because of the need for C1 to recharge through R4. This can be overcome by the addition of a diode and a resistor, as discussed in Chapter 3. When pnp transistors are used in all of the circuits, the supply voltage and all diodes are reversed, and all waveforms will be inverted.

SIMPLE MONOSTABLE

For noncritical applications, where the required pulse is shorter than the input pulse and high accuracy is not required, the simplified circuit shown in Fig. 4-5 may be used. Of course, C1 and R1 are the timing components, and the pulse width produced depends on the amplitude of the input signal. The input should not exceed 5V p-p. If the input signal amplitude is equal to the supply voltage, t equals 0.7C1R1.

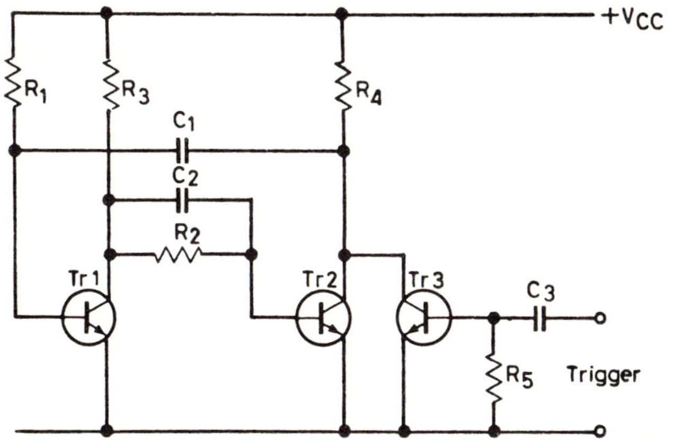

Typical values as Fig. 4-1 except:

R5 = 15K
C3 = 100 pF
Tr3 = ZTX300

Fig. 4-4. Transistor-triggered monostable multivibrator.

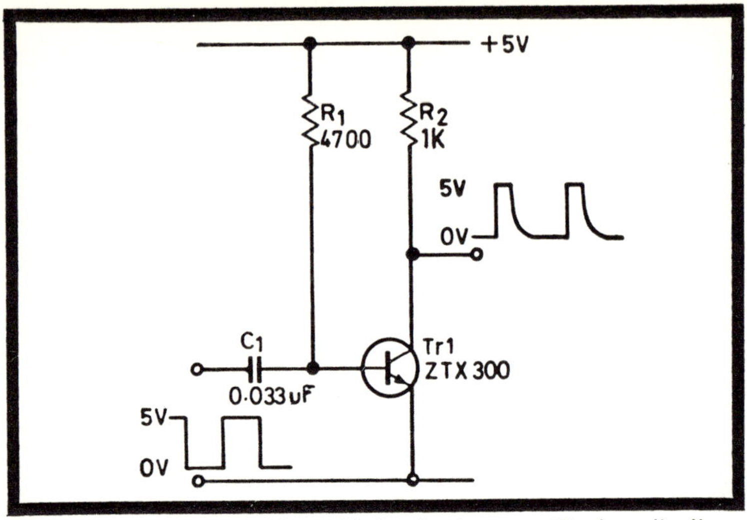

Fig. 4-5. Simple monostable multivibrator for noncritical applications. With values shown, t equals 100 usec.

EMITTER COUPLED MONOSTABLE

In the current-mode circuit of Fig. 4-6, the transistors do not saturate. Under quiescent conditions, Tr1 is cut off and Tr2 is fully conducting. Transistor Tr3 is a constant-current source. The input pulse is differentiated by C2 and R2; on the positive edge, D1 conducts and forward biases the base of Tr1. Then Tr1 starts to conduct, its collector voltage drops, and this change is coupled back via Tr2 and C1 to the emitter of Tr1, causing a further increase in current. This regenerative action ends with Tr2 cut off and Tr1 conducting all of the current provided by Tr3. During the quasi-stable state, C1 discharges linearly, with current defined by Tr3, until the emitter voltage of Tr2 falls to a level at which Tr2 starts to conduct. The current flowing in Tr1 falls, its collector voltage rises, and the circuit rapidly switches back to its original state, charging C1 through Tr2, D2, and R4.

The pulse width can be estimated as follows:

Current provided by Tr3 equals 5.3 mA.

Voltage swing across R1=IR1=5.3 x 750=4V. However, C1 only has to charge through 2.6V because of the "step" produced by the V_{be} of Tr1 and the forward voltage of D2. That is, C1 charging voltage=IR1 - $V_{be1} - V_{d2}$.

IT=CV; therefore, t=

$$\frac{2000 \times 10^{-12} \times 2.6}{5.3 \times 10^{-3}}$$

= 0.98 usec with component values given.

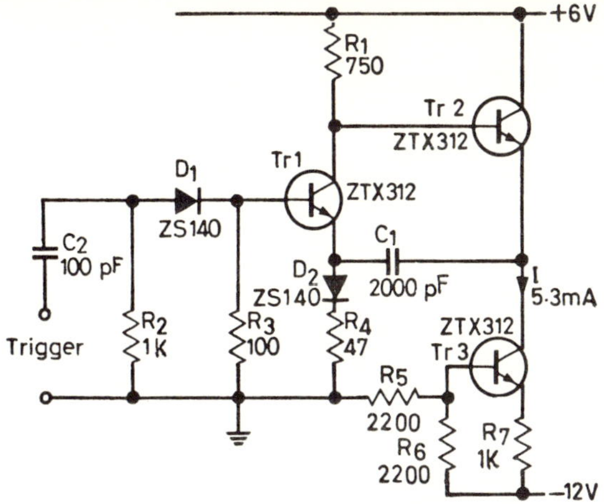

With component values shown, pulse width T=1 usec.
Rise and fall times are less than 20 nsec.

Upper trace	—	Tr1 collector
Lower trace	—	Tr2 emitter
Vertical scale	—	2V/cm
Horizontal scale	—	200 nsec/cm

Trigger—4V p-p minimum. Rise time=10 nsec.

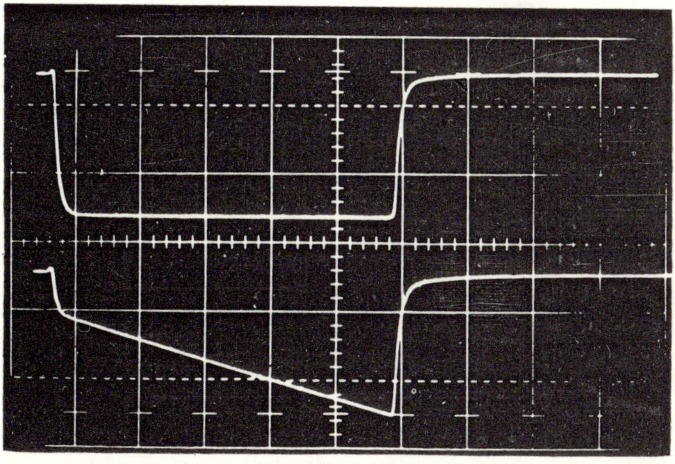

Fig. 4-6. Emitter-coupled monostable multivibrator, with waveforms. With values shown, pulse width, t, equals 1 usec.

Other pulse widths may be obtained by choice of a suitable capacitor. Figure 4-6 shows the actual output waveform obtained, with a measured pulse width of just over 1 usec.

DESIGN EXAMPLE 1

Design a general-purpose monostable to operate from a 5V plus or minus 0.5V supply, with an output impedance of 1K. Resistor tolerance when new will be 10 percent nominal, and assumed 20 percent at end of life. Temperature limits are minus 10 to plus 50 deg C.

With reference to Fig. 4-1, perform the following exercise:

1. Choose suitable transistor (ZTX300).
2. Choose collector load resistors R3, R4; 1K nominal.
3. Minimum supply voltage and minimum temperature are the worst conditions for turning the transistor fully on. Therefore, with 4.5V dc supply, the collector current of Tr1 and Tr2 equals 4.5 mA nominal, and 5.6 mA maximum.
4. Minimum current gain of ZTX300 at 5.6 mA is 45 at 25 deg and 34 at minus 10 deg C.

Hence, the minimum base current to saturate equals 5.6 divided by 34, or 0.165 mA.

5. Maximum base circuit resistance is found from

$$\frac{4.5}{0.165} \times \frac{100K}{120} = 18.7K.$$

That is, R1 max. = 18.7K and R2 max = 17.7K because R3 is in series with R2.

The 15K value was chosen for both R1 and R2 because, with this value, the pulse width produced is numerically similar to the value of C1; i.e., 0.1 uF gives 1 msec.

The worst conditions for triggering in Tr1 is that R1 should be at the low end of its tolerance and R3 should be high. Under these conditions, with a 5.5V supply, the Tr1 base current equals 400 uA. The Tr1 collector current equals 4.6 mA; i.e., I_c divided by I_b is approximately equal to 10.

Putting these figures into the trigger charge graph, Fig. 1-5, gives a trigger charge of 560 picocoulombs. If we assume that the trigger voltage is 4V, then

$$C3 = \frac{560}{4} = 140 \text{ pF}$$

The nearest preferred (standard) value above this is 150 pF, and this indicates a minimum value of 1500 pF for C1, giving a minimum pulse width of 15 usec. Shorter pulses can be obtained with reduced accuracy.

Similarly, for Tr2, but using the stored charge graph of Fig. 1-3, we can see that stored charge equals 390 picocoulombs. With a 5.5V supply, the voltage swing on Tr1 collector is approximately 5V. Hence C2's value is 390/5, or 78 pF. But 100 pF is close enough.

In the finished circuit (Fig. 4-1), if ZTX311 transistors are used instead of ZTX300, C3 can be reduced to 10 pF and C2 to 5 pF minimum. This will produce a minimum pulse width of 1 usec, with 100 pF at C1.

DESIGN EXAMPLE 2

Design a high-speed monostable with pulse width variable over a 4 to 1 range, down to 100 usec.

1. To achieve maximum speed, it is desirable that a low value of load resistor should be used, and that the transistors should be operated close to the peaks of their h_{fe}/I_t and f_t/I_c curves. With ZTX312 transistors, this occurs in the 10 to 20 mA region. To obtain the best possible waveform, a 12V supply is used, with catching diodes limiting the collector voltage to 5V. An isolating diode (D3) is used to provide a separate path for timing capacitor C.

Choosing a value of 1K for R5, R6, the nominal collector current of Tr2 will be 12 divided by 500, or 24 mA.
The maximum collector current of Tr2=

$$24 \times \frac{100}{80}$$

Minimum current gain of Tr2 at 30 mA is 35 at 25 deg C and 29 at 0 deg C.

Hence, minimum base current to saturate=

$$\frac{30}{29} = 1.03 \text{ mA}$$

2. The nominal base current= $\frac{1.03 \times 120}{100} = 1.24$ mA.

Since Tr2 collector is caught at 5V, the maximum value of R4=

$\frac{5 - 0.85}{1.24}$=3.35K, that is, 3.3K nominal.

3. A value of 1K was chosen for R3 to provide adequate drive to Tr2 and to allow Tr1 to operate close to the peaks of h_{fe}/I_c and f_t/I_c curves. It can be seen that maximum current required from Tr1=

$$\frac{12}{1K} \times \frac{100}{80} = 15 \text{ mA.}$$

4. Current gain of Tr1 at 15 mA is 40 at 25 deg C, and 33 at 0 deg.

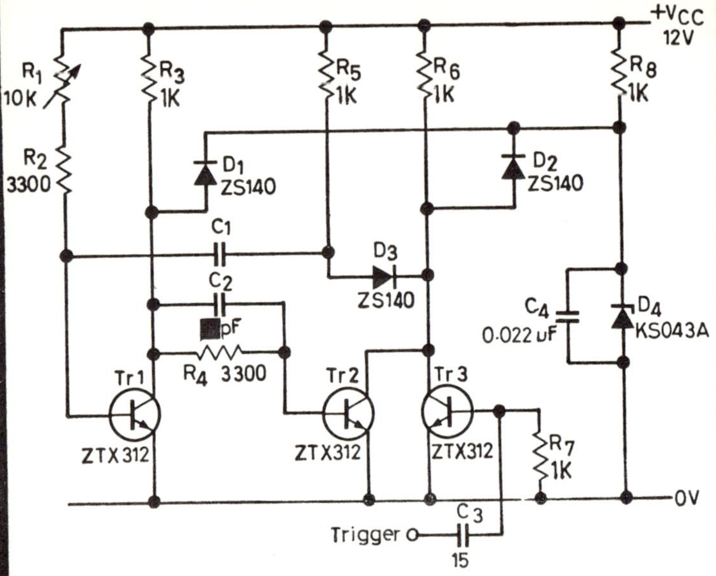

Upper trace — Tr1 collector
Lower trace — Tr2/Tr3 collectors
Vertical scale — 2V/cm
Horizontal scale — 100 nsec/cm

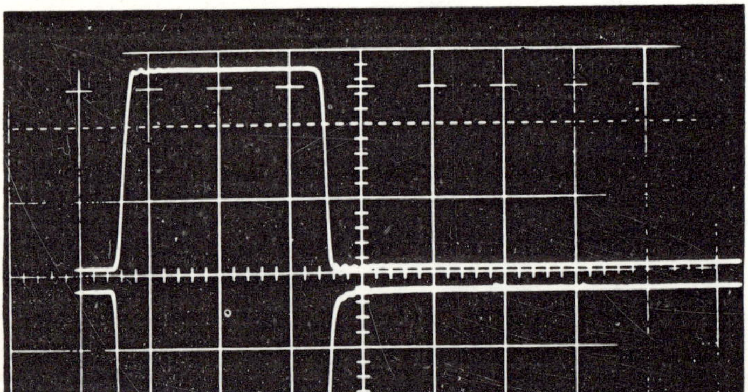

Fig. 4-7. Fast monostable multivibrator and output waveforms with 68 pF value of C1.

Minimum base current for saturation=

$$\frac{15}{33} = 455 \text{ uA.}$$

Maximum base resistance R1 + R2=

$$\frac{(12 - 0.85)}{0.455} \times \frac{100\text{ K}}{120} = 20.4\text{K.}$$

A lower value than this can be used and, in order to obtain a 4 to 1 range, R1 was selected to be 10K and R2 to be 3.3K. This completes the dc design.

5. With R5 and R6 high, and R4 low, the collector current of Tr2 is 20 mA, and the base current of Tr2 is 1.9 mA, because I_c over I_b is approximately 10.

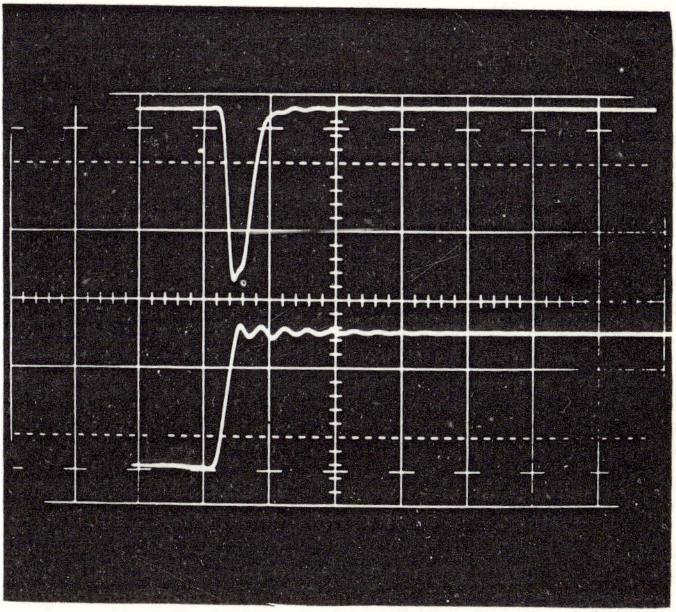

Upper trace	— Tr2/Tr3 collectors
Lower trace	— Trigger waveform
Horizontal scale	— 50 nsec/cm
Vertical scale	— 2V/cm

Fig. 4-8. Trigger pulse at collectors of Tr2-Tr3 of Fig. 4-7 with C1 disconnected.

Referring to the store charge graph, Fig. 1-3, the store charge under these conditions is 50 picocoulombs. With a 5V swing on Tr1 collector, the minimum value of C2 is 50 divided by 5, or 10 pF.

6. Transistor triggering was chosen because this gives high trigger sensitivity with good isolation between the timing circuit and the trigger source. Resistor R7 was chosen empirically as 1K, and C3 was increased in value until reliable triggering was obtained under all conditions. With a 4V p-p trigger signal, this occurred with C3 at 10 pF. A value of 15 pF has been chosen for the final circuit to allow for component spreads. Because R7C3 is 15 nsec, the trigger pulse is much shorter than the minimum pulse width required, and will not affect the timing.

7. Capacitor C1 is chosen according to the pulse width required. The timing equation is $t = 0.38C1\ (R1 + R2)$. The output waveforms with C1 at 68 pF and R1 adjusted for a pulse width of 300 nsec is shown in Fig. 4-7. Rise and fall times are approximately 10 usec. Figure 4-8 shows the trigger pulse at Tr2 collector with C1 removed.

Chapter 5

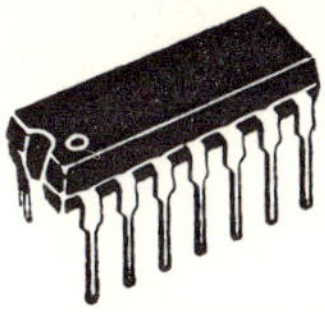

Bistable Multivibrators

If we replace the two ac couplings of the basic astable multivibrator with two dc couplings, we get a bistable multivibrator (Fig. 5-1). This circuit has two stable states and will remain in either until triggered into the opposite state by an external pulse. Bistables have many applications, including counters, shift registers, and memory elements. It is usually more economical to use integrated circuits than discrete components. The exceptions are: (a) when the supply voltage or the input and output requirements are incompatible with integrated circuits; (b) when very low power dissipation is required; (c) when very low or very high speed operation is required; and (d) when only a very simple counting circuit is required (as, for example, in electronic organ dividers).

Counting bistables are the most suitable for discrete components, so only this type will be described.

BASIC COUNTER CIRCUIT

Assume that initially Tr1 is saturated and Tr2 is cut off (Fig. 5-1). Under these conditions, D2 is reverse biased by slightly less than Vcc but D1 has a small forward bias. The trigger pulse is differentiated by C3R5 and C4R6, and, on the negative edge of the waveform, D1 conducts and reverse biases Tr1. As Tr1 starts to turn off, its collector voltage rises. Transistor Tr2 starts to conduct and its collector voltage falls, reducing the base current of Tr1. The regenerative action ends with Tr2 saturated and Tr1 cut off. The bias on the steering diodes, D1 and D2, is now reversed so that the negative trigger pulse cuts off Tr2 and the circuit reverts to its original state. For high-speed operation, D3 and D4 can be added, if desired, to reduce the recharge time of the trigger capacitors.

The method of triggering shown in Fig. 5-1 is the most popular and generally the most satisfactory, but there are other methods. Figure 5-2 shows an example of collector triggering. Assuming that Tr1 is saturated and that Tr2 is cut off, D1 will be reverse biased by 5V and D2 by about 0.5V. The trigger pulse is differentiated by C3 and R5, and, on the negative edge of the waveform, D2 conducts, pulling down the collector voltage of Tr2. This transient change is coupled via

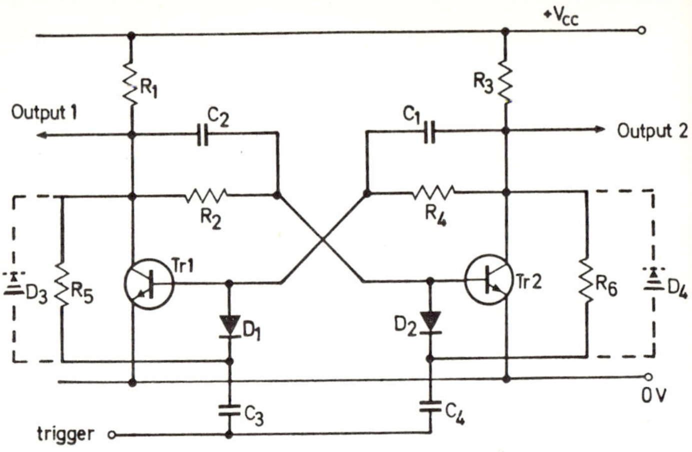

Typical component values:

R1	=	R3	=	1K
R2	=	R4	=	10K
R5	=	R6	=	10K
C1	=	C2	=	220 pF
C3	=	C4	=	470 pF
D1	=	D2	=	ZS140
D3	=	D4	=	ZS140
Tr1	=	Tr2	=	ZTX300
$+V_{cc}$	=	5V		

Note: Without D3, D4 maximum trigger rate = 200 kHz
With D3, D4 maximum trigger rate = 400 kHz

Fig. 5-1. Basic bistable multivibrator.

C1 to the base of Tr1, causing Tr1 to turn off. As the Tr1 collector voltage rises, Tr2 turns on, its collector voltage falls further, and the circuit switches.

Because C1 and C2 have to act as a transient memory, their value must be greater than would be suggested by charge-storage considerations. For the ZTX300 transistors, the minimum for C1, C2 is 470 pF in the circuit shown, but higher values may be used. Problems sometimes arise with this circuit when several stages are cascaded because there is no isolation between the trigger and the output. The maximum trigger rate is 100 kHz. With the component values shown, the circuit will operate at trigger rates up to 50 kHz, but the output waveform is poor above 20 kHz.

Figure 5-3 shows a very simple method of triggering suitable for low-speed bistables. The trigger waveform is differentiated by C3R5. If Tr1 is saturated and Tr2 is cut off, nothing appears on the Tr1 collector, but the differentiated trigger waveform appears on the collector of Tr2. The negative pulse edge is coupled via C1 to the base of Tr1, and momentarily turns it off. As Tr1 collector voltage rises, Tr2 starts to turn and rapidly turns off Tr1. The state of the circuit reverses again at the next negative trigger pulse.

The capacitor values shown in Fig. 5-3 are the minimum for the component values used, and this circuit will operate at trigger rates up to 50 kHz. For operation at lower speeds larger values of capacitance are desirable; however, a ratio of 2 to 1 should be kept between C3 and C1-C2.

DETERMINING MINIMUM VALUES OF TRIGGER AND SPEEDUP CAPACITORS

Taking the circuit shown in Fig. 5-1 as an example, the minimum value of speedup capacitance can be determined as follows. Assuming 20 percent resistor life tolerance and 10 percent supply voltage tolerance, the minimum collector current with a 5.5V supply=

$$\frac{5.5}{1000} \times \frac{100}{80} = 6.87\,\text{mA}.$$

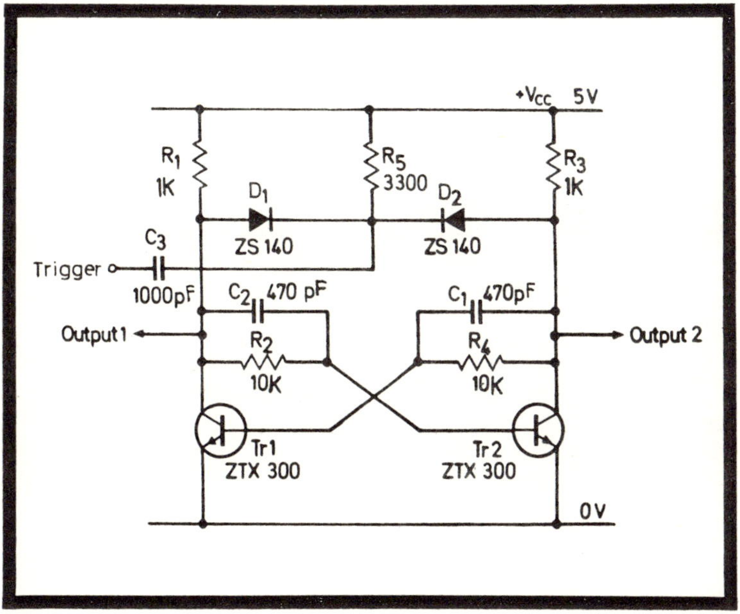

Fig. 5-2. An example of collector triggering. Maximum trigger rate is 100 kHz.

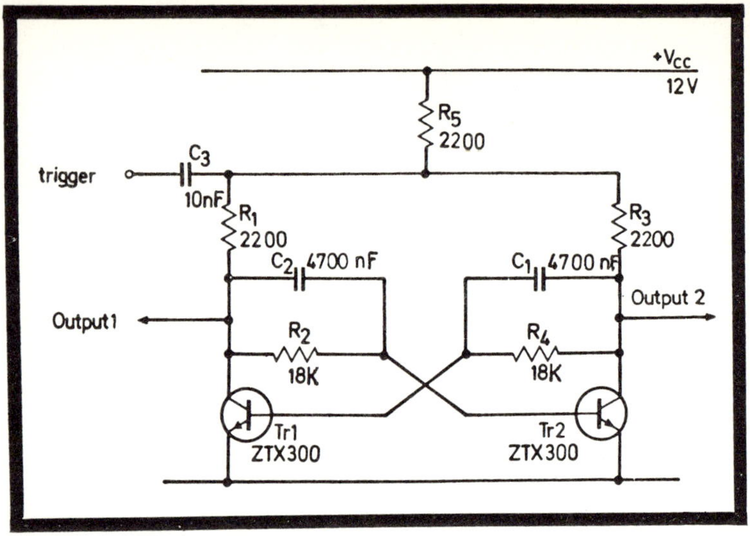

Fig. 5-3. A simple method of collector triggering for low-speed biastable multivibrator.

The maximum base current with a 5.5V supply =

$$\frac{(5.5-0.7)}{11K} \times \frac{100}{80}$$

$= 0.55$ mA. That is,

$$\frac{I_c}{I_b} = 8.4$$

From the speedup charge graph, Fig. 1-3, we can see that the stored charge of a ZTX300 under these conditions is approximately 550 picocoulombs. Assuming a collector voltage swing of 5V p-p, the minimum speedup capacitance is 110 pF.

Putting the above figures into the trigger charge graph, Fig. 1-5 indicates a minimum trigger charge of about 700 picocoulombs—that is, 175 pF minimum with a 4V p-p minimum trigger signal. A slightly higher value than this is required in practice, however, because some of the trigger charge is lost in steering diodes D1 and D2 and in steering resistors R5 and R6. The true value may be determined with the circuit shown in Fig. 5-4, which is, in effect, half of the bistable built with worst-case component values to ensure that the transistor is heavily saturated.

It was found that the minimum value of trigger input capacitor C to fully turn off all samples of ZTX300 was 220 pF.

To ensure reliable operation, larger values of capacitance have been used in the final circuit, the speedup capacitors being 220 pF and the trigger capacitors 470 pF. The dc design for this bistable follows the same lines as those for astables and monostables, and will not be described here.

BINARY COUNTERS

Any number of bistables may be connected in cascade, the count being 2n where n is the number of stages. When division ratios that are not a power of 2 are required, feedback, feed-forward, and gating can be used to obtain the required count.

Figure 5-5 shows a 2-stage counter modified to give a count of 3 by means of feedback from Tr3 and Tr2. Assuming that Tr2 and Tr4 are initially in the "0" (low) state (conducting with V_{ce}=0.2V), the first input pulse triggers Tr2 into the "1" (high) state. The second input pulse returns Tr2 to the "0" state, which in turn triggers Tr4 into the "1" stage. Transistor Tr3, meanwhile, goes to "0" and triggers Tr2 back into the "1" state (all of this happens in less than 1 usec). The third input pulse returns Tr2 to the "0" state, which in turn returns Tr2 to the "0" state, and the circuit reverts to its starting point.

A thorough study of the circuit, block diagram, and especially the truth table and waveforms, all shown in Fig. 5-5, is encouraged before you proceed.

Feedback from the last to the first stage of a 3-stage binary counter would give a count of 7 instead of 8; to the second stage would give a count of 6 (2 x 3); and to both the first and second stages would give a count of 5. In general, feedback to the first stage of a counter will reduce the count by 1, to the second stage by 2, to the third stage by 4, etc. with any

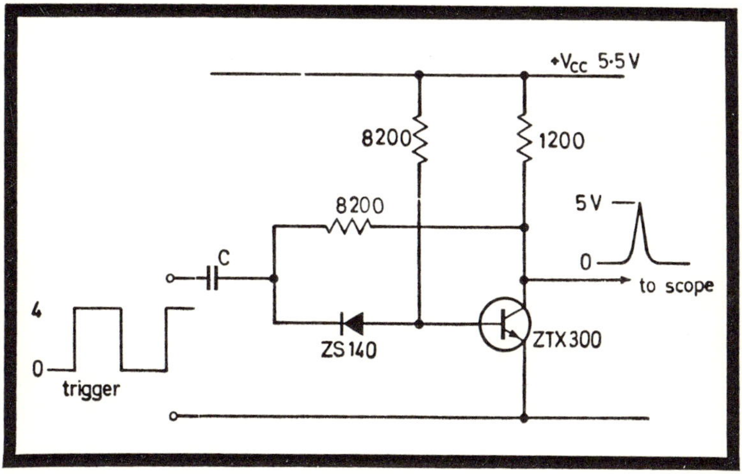

Fig. 5-4. Test circuit for determining the minimum value of trigger capacitance needed for the circuit of Fig. 5-1. Input capacitor C is increased until an output pulse of 5V minimum amplitude is obtained with all transistor samples of a single type.

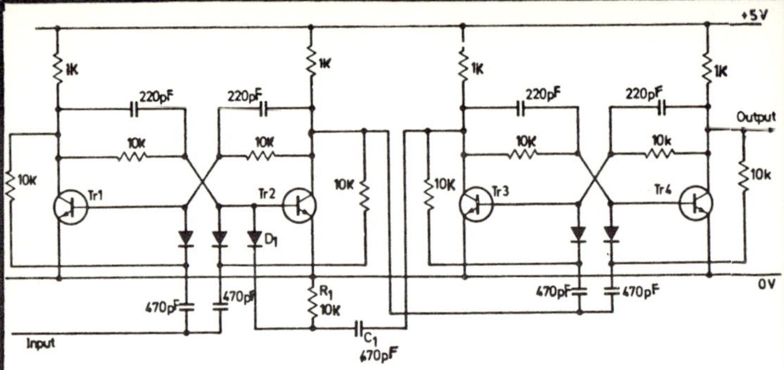

All transistors — ZTX300
All diodes — ZS140

Pulse No.	Tr2	Tr4
0	0	0
1	1	0
2	0 transient state	1
2	1	1
3	0	0

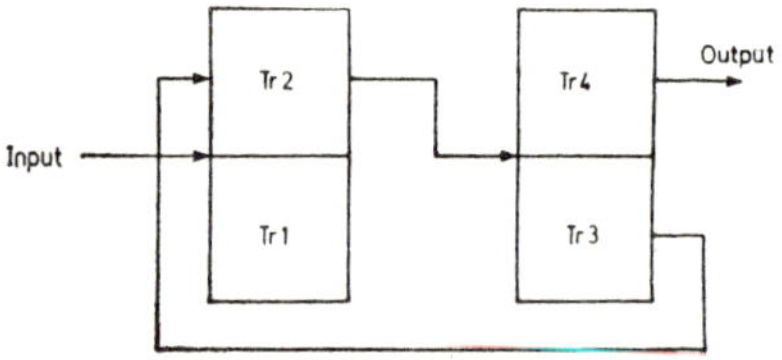

0 = low (0.2V), 1 = high (4.5V)

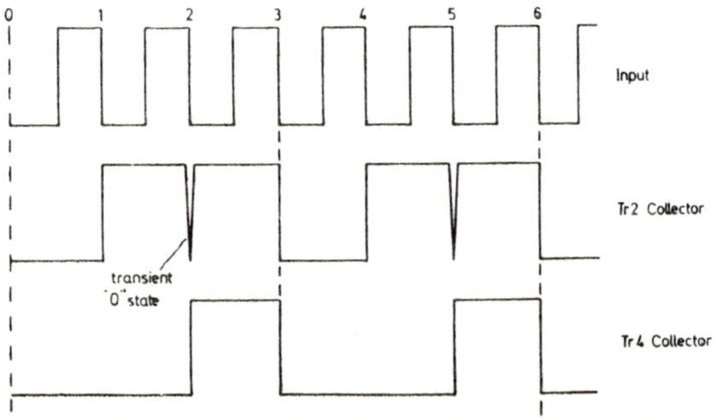

Fig. 5-5. A 2-stage binary counter with feedback to reduce the count to 3.

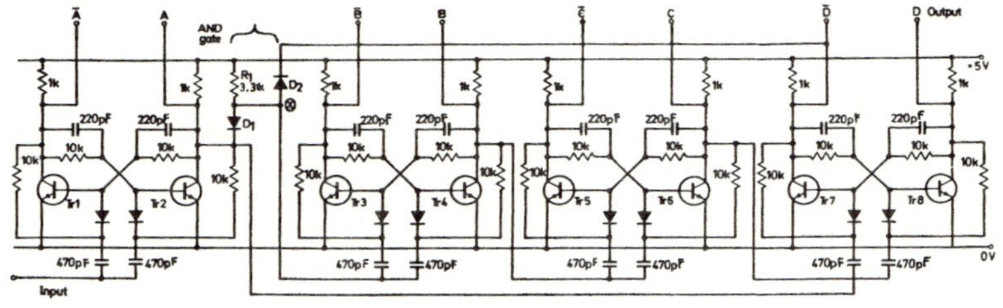

All transistors — ZTX300
All diodes — ZS140

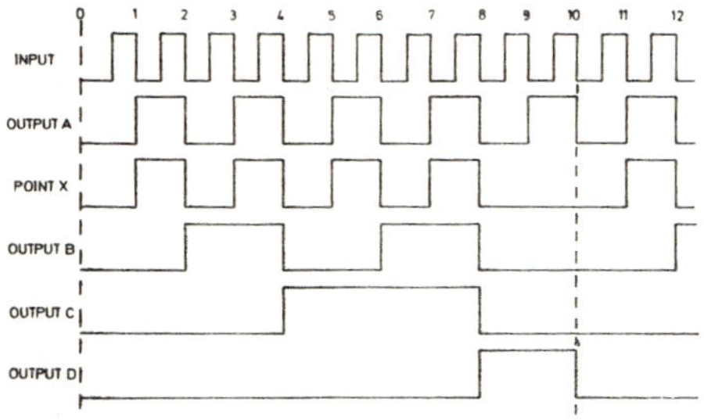

Waveforms

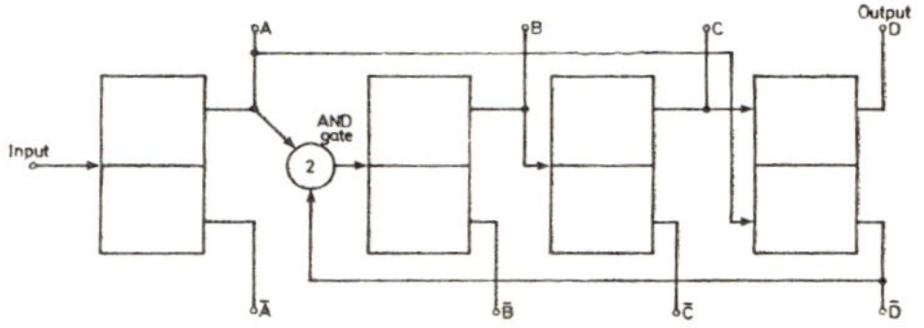

Schematic

Pulse No.	D	C	B	A	X
0	0	0	0	0	0
1	0	0	0	1	1
2	0	0	1	0	0
3	0	0	1	1	1
4	0	1	0	0	0
5	0	1	0	1	1
6	0	1	1	0	0
7	0	1	1	1	1
8	1	0	0	0	0
9	1	0	0	1	0
10	0	0	0	0	0

Fig. 5-6. Decade counter using four bistable multivibrators, with an AND gate between the first and second stages and feed-forward from the first stage to reset the fourth stage.

combination being possible. This technique is quite satisfactory for simple frequency dividers and counters; where all of the outputs are used, however, take care to ensure that the transient "0" states produced do not interfere with other circuits.

Figure 5-6 shows a decade counter, using gating between the first and second stages and feed-forward from the first to the fourth stage. Up to the eighth pulse, the AND gate is open, and the circuit operates as a simple binary counter. On the eighth pulse, Tr8 is triggered from the "0" state to the "1" state while Tr7 goes to "0", closing the AND gate.

The ninth pulse triggers Tr2 into the "1" state. On the tenth pulse, Tr2 goes to "0" and resets Tr7 to the "1" state and

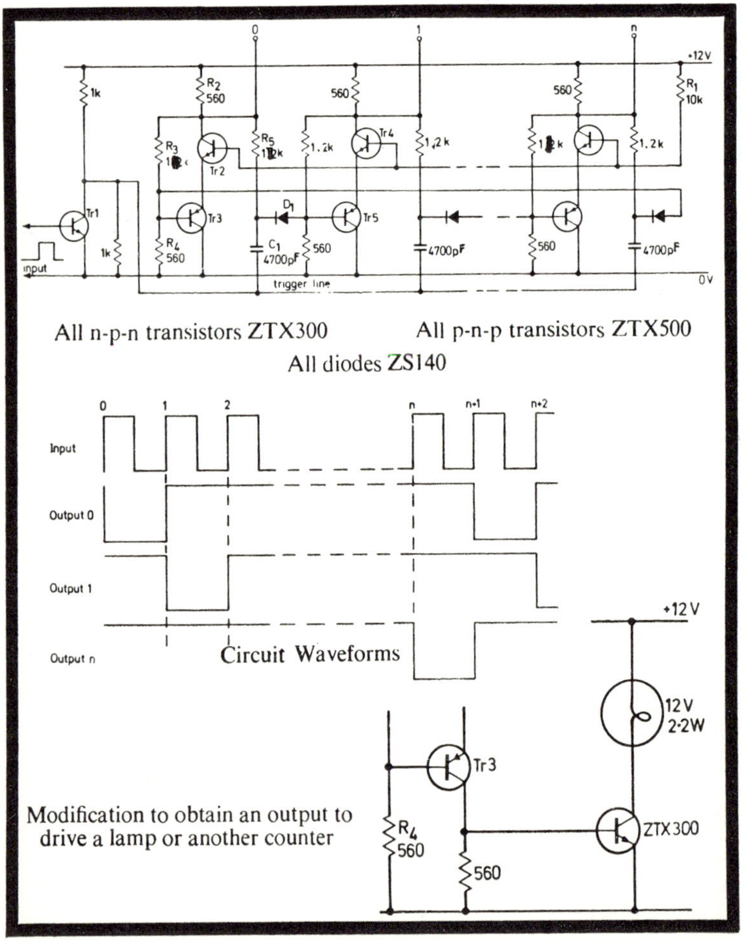

Fig. 5-7. Practical ring-counter circuit. (All stages are identical, and any number can be connected in cascade.)

Tr8 to "0." The cycle then repeats. No transient states are produced by this circuit.

RING COUNTERS

Ring counters are useful when you require a count that is not a power of 2 or when you require a separate output for each number instead of a binary coded output. Realize, however, that ring counters tend to become uneconomical at counts above 10.

Figure 5-7 shows a practical circuit. Each of the npn-pnp pairs forms a bistable element, although bistable action is only exhibited when two or more stages are connected. Only one stage can conduct at any given time, and the conducting stage takes all of the base current provided by R1 and steers the trigger pulse to the next stage in the ring. Assume that the "0" stage is conducting and all other stages are cut off. Then Tr2 is saturated with about 1.5V on its collector and 2.2V on its base. The base voltage of Tr3 is about 0.5V. For all other stages, the collector and base of the npn will be at 9V and 2.2V, respectively, while the base of each pnp will be at 2.7V, i.e. cut off. Diode D1 is forward biased by about 1.2V and all other diodes are reverse biased by about 7V. The first negative trigger pulse drives the base of Tr5 negative; it starts to conduct and the collector voltage of Tr4 falls, causing regenerative action that ends with the "1" stage conducting and all of the other stages cut off. The process repeats at each trigger pulse.

Any number of stages may be connected in the ring. The output of the last stage is always fed back to the input of the first. A direct readout of the state of the counter can be obtained if the collector load resistor (R2 of each stage) is replaced with a 6V, 40 mA lamp in series with a 100 ohm resistor. Transistors ZTX302 and ZTX502 are preferred in this case. To drive a more powerful lamp or to obtain an output to drive another counter, the output circuit shown at bottom right of Fig. 5-7 may be used.

Chapter 6

Digital Display Circuits, and Digital-to-Analog Converters

It is frequently necessary to present a numerical display for such applications as digital voltmeters, frequency meters, summation metering, etc. The modern trend is also to replace the moving-coil instrument by digital readout. This may be achieved by a variety of indicators. The circuits described below show a few of these, along with the associated driving components.

COUNTER

Figure 6-1 shows a four-stage binary counter, using Micronor 2 elements, suitable for use with the Nixie driver circuit. Similar circuits can be built using TTL elements, high-level logic or with discrete components, as shown in Fig. 5-6.

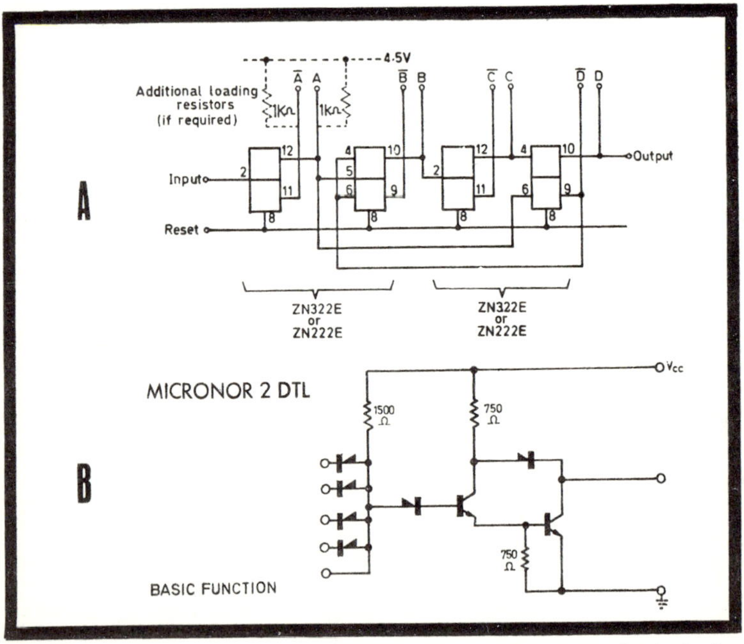

Fig. 6-1. Counter (A) using Ferranti Micronor 2 DTL elements. In B, the DTL schematic.

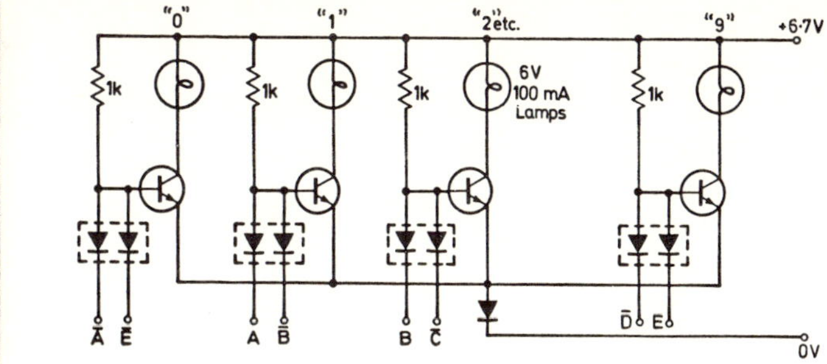

All diode pairs
ZDX1R

All transistors
ZTX300

Pulse No.	Outputs A	B	C	D	E	Decoding logic
0	0	0	0	0	0	$\bar{A}.\bar{E}$
1	1	0	0	0	0	$A.\bar{B}$
2	1	1	0	0	0	$B.\bar{C}$
3	1	1	1	0	0	$C.\bar{D}$
4	1	1	1	1	0	$D.\bar{E}$
5	1	1	1	1	1	$A.E$
6	0	1	1	1	1	$\bar{A}.B$
7	0	0	1	1	1	$\bar{B}.C$
8	0	0	0	1	1	$\bar{C}.D$
9	0	0	0	0	1	$\bar{D}.E$

Positive logic, 1 = high
0 = low

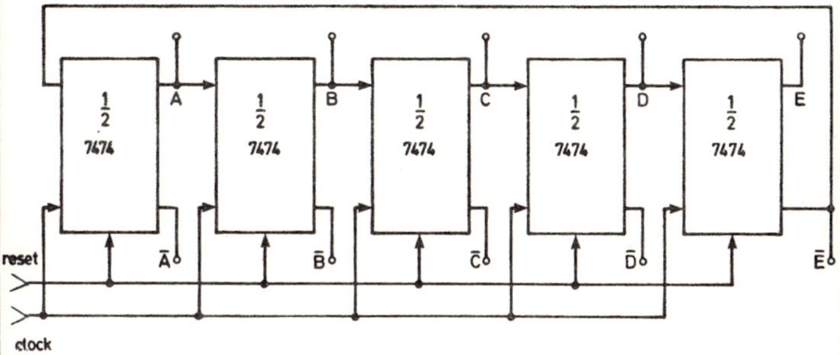

Fig. 6-2. Filament-lamp readout using ZDX1R diode paris and ZN7474 (SN7474) dual D-type flip-flop in a Johnson decade counter.

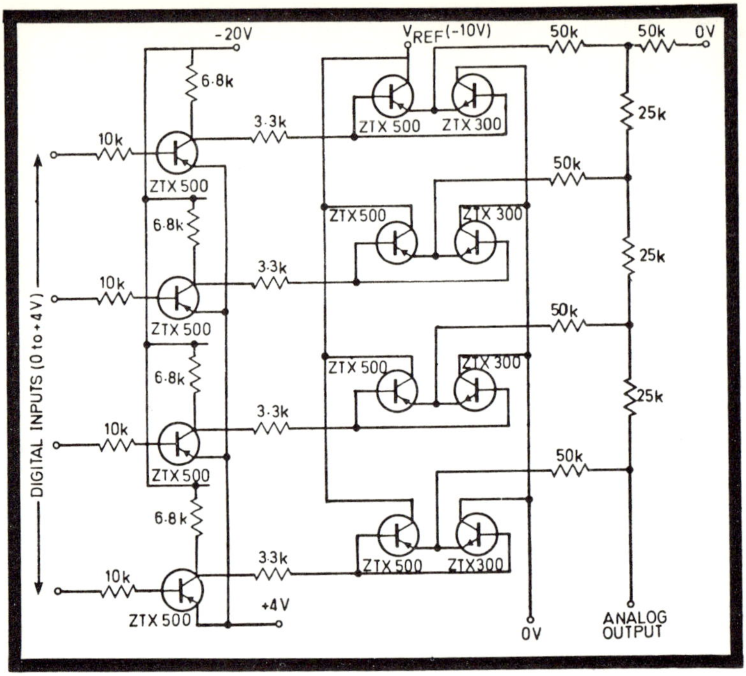

Fig. 6-3. Digital-to-analog converter.

PROJECTION AND EDGE-ILLUMINATION READOUT

This type of indicator uses incandescent lamps, usually 10 per unit. The driving circuit must be able to energize any one of the lamps according to the logic states present at the input. A suitable circuit driven by a Johnson decade counter with ZN747 D-type flip-flops is shown in Fig. 6-2. This type of counter requires only a simple 2-input AND gate to give a unique output for each number, and ZDX1R diode pairs can be used to advantage. The common-emitter-circuit diode provides a dc bias which holds off all except the selected transistor. Component values shown are suitable for lamps of 6V, 100 mA.

4-BIT DIGITAL TO ANALOG CONVERTER

Figure 6-3 illustrates the use of complementary transistors as switches. Driver transistors are used, operating from supply buses having voltage greater than the switch supply buses to ensure that the switching transistors will "bottom" satisfactorily. A ladder network, consisting of suitable close-tolerance resistors, is used to sum the components.

Chapter 7

Ramp & Staircase Generators and Pulse Counter

When a capacitor C is charged from 0V toward a voltage V, and C is in series with a voltage V, and C is in series with a resistor R, the voltage across the capacitor rises exponentially according to

$$v = V \left[I - \exp(-t/RC) \right]$$

where v is instantaneous voltage and t is time.

In most ramp-generator applications, a **linear** charge is required, and various methods of overcoming the nonlinearity of the RC circuit have been devised.

CONSTANT CURRENT RAMP GENERATOR

In the method shown in Fig. 7-1, the high output impedance of transistor Tr2 is used to give a close approximation to constant-current charging, with the current being defined by R1, R2, and R3. Transistor Tr3 provides a low-impedance output, and component values for this stage must be chosen so that the base current of Tr3 is much less than the charging current. In this circuit, Tr3 runs at 2 mA and requires a base current of 20 uA maximum (10 uA typical); for nonlinearity to be negligible, the charging current should be about 1 mA.

For a ramp waveform of 1 msec duration and 5V amplitude, CV equals IT. Therefore,

$$C = \frac{tI}{V}$$

$$= \frac{10^{-3} \times 10^{-3}}{5}$$

$$= 0.2 \text{ uF}.$$

The waveforms obtained are as shown.

Figure 7-2 shows how the circuit can be modified so as to be self-oscillating. The output waveform obtained is virtually the same as for Fig. 7-1. Timing capacitor C charges until its voltage exceeds the base voltage of Tr4 by 0.7V. Then Tr4 conducts and, as Tr1 starts to conduct, the base voltage of Tr4

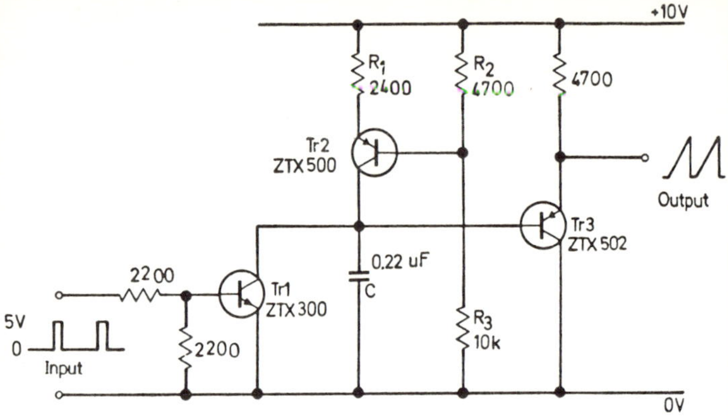

Upper trace — output (2V/cm)
Lower trace — input (5V/cm)
Horizontal scale — 200 usec/cm

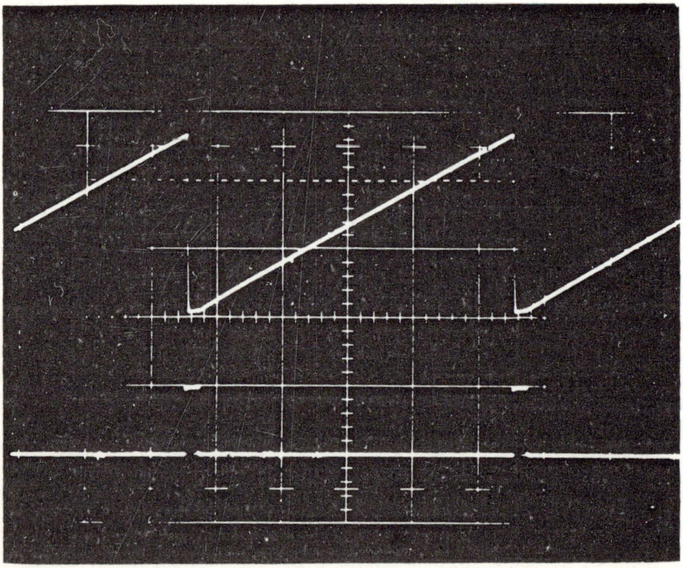

Fig. 7-1. Constant-current ramp generator.

falls, providing regenerative feedback. Both Tr1 and Tr4 saturate and rapidly discharge the capacitor to approximately 0.7V. Diode D1 cuts off the current source so that, as soon as C has been discharged, Tr1 and Tr4 turn off and the linear charge starts again. With the transistors shown, ramp times down to 10 usec can be obtained by choice of a suitable capacitor. If Tr1 and Tr4 are replaced with ZTX310 and ZTX510, respectively, ramp times down to 1 usec can be obtained.

BOOTSTRAP RAMP GENERATOR

As is well known, if the voltages at both ends of a resistor change by the same amount, no change in current will occur. This is basically what the bootstrap circuit does. Figure 7-3 shows a practical circuit and the output waveforms obtained. Here, R1 and R2 replace the current generator of Fig. 7-1, and provide a charging current of 1 mA. Resistor R3 stops parasitic oscillations.

Parabolic Correction

If a sawtooth waveform is integrated with an RC network, a parabolic waveform is obtained. If the parabolic waveform is then added to an exponential ramp, in the correct proportions, the nonlinearity of the ramp can be canceled.

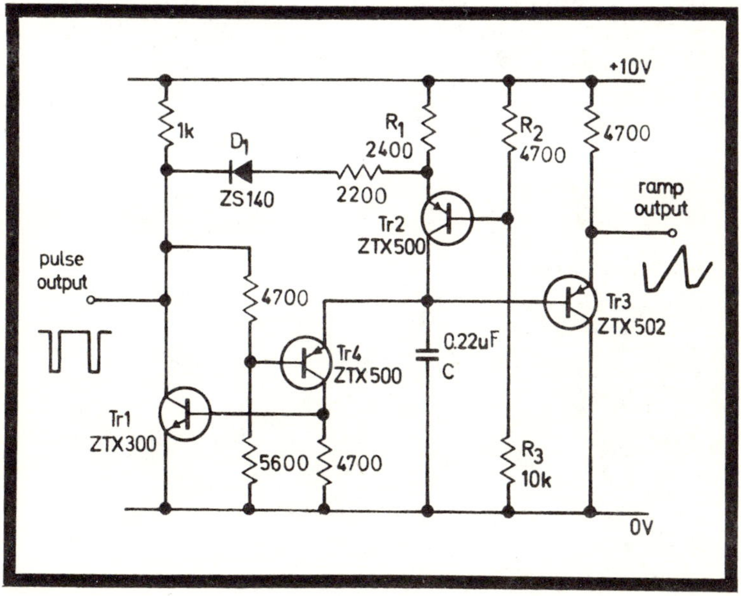

Fig. 7-2. A self-oscillating version of Fig. 7-1. A synchronizing signal can be applied to the base of Tr4.

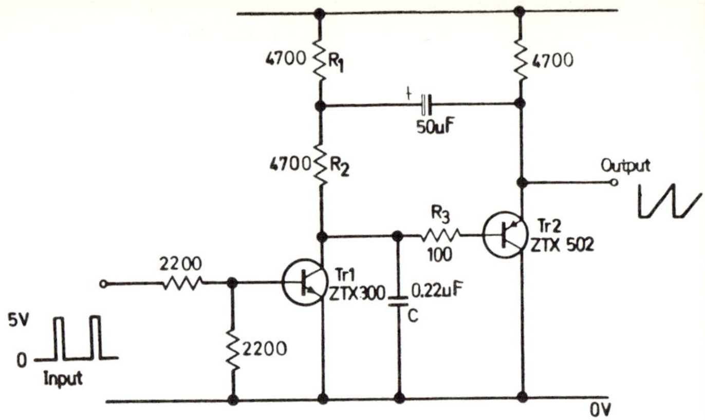

Upper trace —with bootstrap capacitor (2V/cm)
Lower trace —without bootstrap capacitor (2V/cm)
Horizontal scale — 200 usec/cm

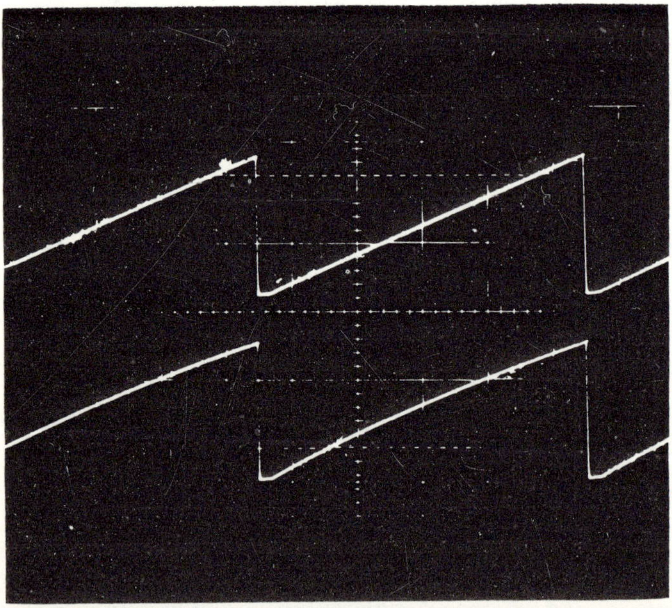

Fig. 7-3. Bootstrap ramp generator.

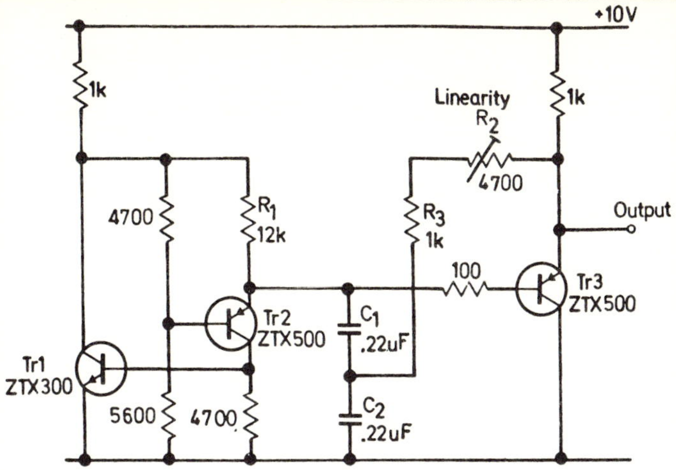

Upper trace	—output with R2 adjusted for best linearity (2V/cm).
Center trace	—waveform across C2 (2V/cm).
Lower trace	—parabolic component of center trace, obtained by cancelling sawtooth component (0.5V/cm).

Horizontal scale — 200 usec/cm

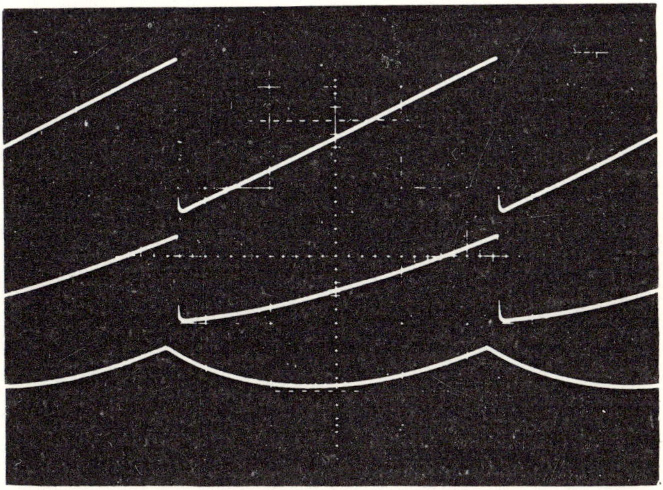

Fig. 7-4. Ramp generator with parabola correction.

Figure 7-4 shows a practical, self-oscillating circuit employing this technique. The charging resistor is R1 and the timing capacitor is split into C1 and C2 (C1 and C2 need not be equal, although this is often convenient). The output is fed back to C2 via R3 and linearity control R2, and the integration is performed by the time constant of the components.

This type of circuit is particularly useful for field-scan generation in TV receivers because the facility of adjustable linearity allows nonlinearity in the field output stage to be corrected.

Feedback Integrator

Assume that initially Tr1 is saturated and Tr2 is cut off. When Tr1 turns off, the current flowing in charging resistor R1 flows into the base of Tr2, which starts to conduct. The collector voltage of Tr2, and the emitter voltage of Tr3 fall, and a current flows through C to offset that provided by R1. Because the base voltage of Tr2 remains virtually constant, the current in R1 remains constant—and capacitor C discharges linearly until the reset pulse arrives. Then Tr1 saturates, Tr2 is cut off, and C recharges current to a safe value. Capacitor C2 is a parasitic stopper.

For good linearity, the base current required by Tr2 should be much less than the charging current. One disadvantage of the circuit shown is that a slight delay occurs before the start of the ramp because of the need to charge C through V_{be} before Tr2 starts to conduct.

Wide-Range Oscilloscope Timebase

The circuit shown in Fig. 7-6 is intended for scope time bases. This circuit will generate a linear ramp of 5V amplitude over a range of 100 msec to 1 usec, and will synchronize up to 30 MHz minimum with 20 mV trigger input.

Transistors Tr1, Tr2, and Tr3 form a trigger amplifier and Schmitt trigger circuit. With S1 in position A (as shown), the bias for Tr1 is derived from the output of the signal, the circuit oscillates continuously at about 50 Hz. This keeps the ramp generator running, provides a baseline on the scope screen, and also biases Tr1 for maximum sensitivity. With an input sine wave of more than 10 mV rms, the circuit triggers normally at signal frequency. With S1 in position B, a dc bias is applied to Tr1 for triggering from low-frequency or nonperiodic waveforms.

Transistors Tr5 and Tr6 form a bistable, and Tr8, Tr9, and Tr10 form a constant-current ramp generator. Assume that initially Tr6 is on, that Tr7 is off, and that reset Tr8 has fully recharged C1 (base of Tr9). The first trigger pulse from the Schmitt trigger circuit cuts off Tr6, the bistable changes state, and Tr8 turns off; this allows C1 to discharge linearly at a rate

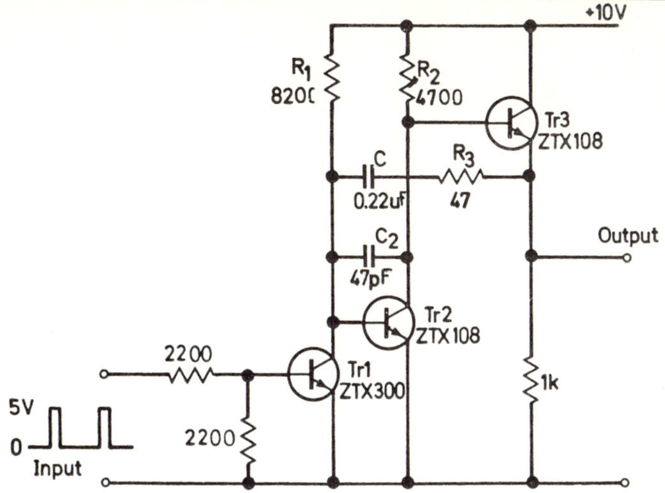

Upper trace —output (2V/cm).

Lower trace —input (5V/cm).

Horizontal scale — 200 usec/cm

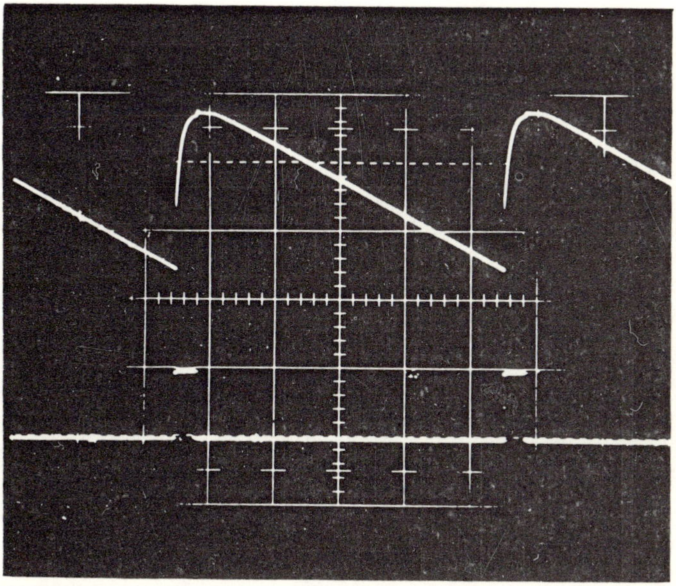

Fig. 7-5. Ramp generator using feedback integrator.

All transistors are ZTX311 unless otherwise specified.

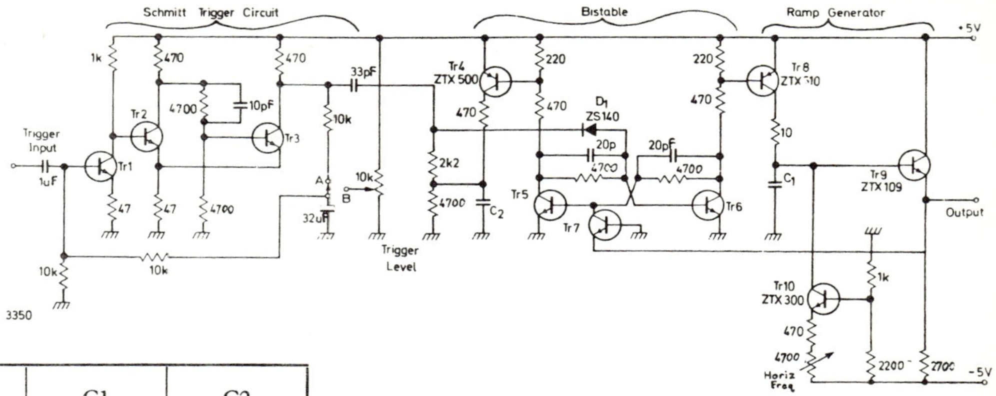

	Speed Range	C1	C2
1	1-10 usec	1000pF	100pF
2	10-100 usec	.01 μF	1000pF
3	100 usec—1 msec	.1 μF	.01 μF
4	1-10 msec	1.0μF	.1 μF
5	10-100 msec	10μF	1.0μF

Note. Values of C1 and C2 were chosen so that only one bank of capacitors is required; i.e., C1 on range 1 becomes C2 on range 2, etc.

Fig. 7-6. Wide-range oscilloscope time-base circuit.

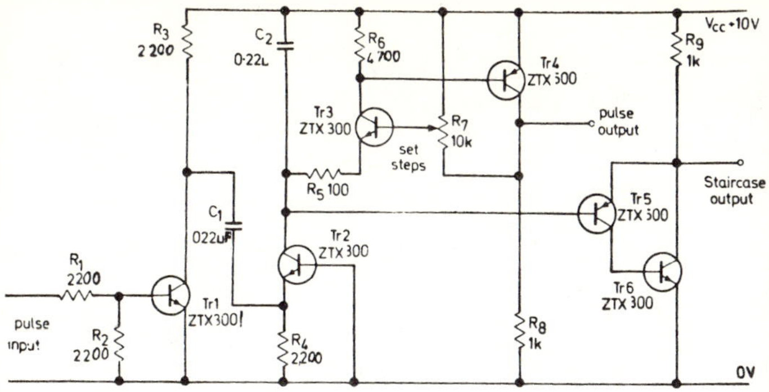

Input frequency range—50Hz-2kHz.

Top trace—output waveform with 4 steps (2V/cm).

Centre trace—input waveform (5V/cm).

Bottom trace—output pulse from Tr4 (5V/cm).

Horizontal scale — 1 msec/cm

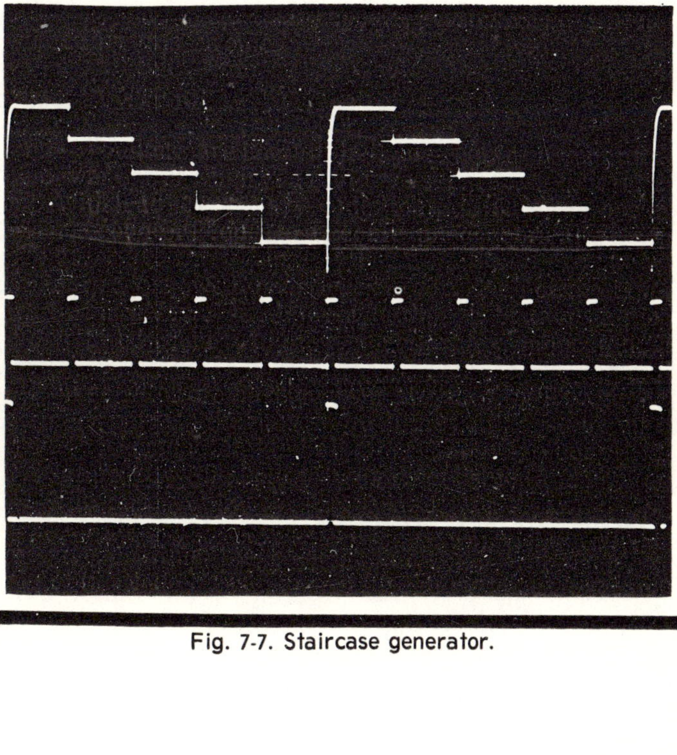

Fig. 7-7. Staircase generator.

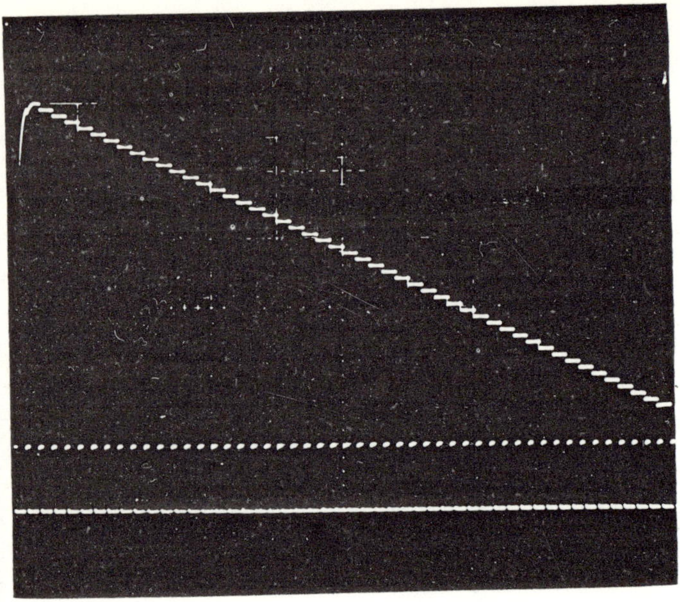

Input frequency —10 kHz
Upper trace —Output 1V/cm
Lower trace —Input 5V/cm
Horizontal scale —500 usec/cm

Fig. 7-8. Output waveform with 50 steps obtained by reducing C1 to 2.2uF.

determined by the setting of HORIZ FREQ potentiometer. At the same time, Tr4 turns on and reverse biases trigger diode D1 so that no further trigger pulses can reach Tr6. When the emitter voltage of Tr9 falls to −0.7V, Tr7 starts to conduct and turns off Tr5. The bistable then resets to its original state and C1 rapidly recharges through Tr8. Meanwhile, Tr4 has turned off, and the voltage across C2 decays until it is low enough for trigger pulses to reach Tr6. The cycle then repeats.

STAIRCASE GENERATORS

A staircase waveform can be obtained by charging a capacitor with current pulses. Figure 7-7 is an example. When Tr1 turns off, C1 charges up to V_{CC} via R3 and R4, and stores charge $Q1 = C1 \times V_{CC}$. When Tr1 turns on, the emitter of Tr2 is forward biased and the charge stored on C1 is transferred to C2. At each pulse, the voltage across C2 increases by approximately

$$\frac{Q1}{C2} = \frac{C1 \times V_{cc}}{C2}$$

Transistors Tr3 and Tr4 discharge C2 every time the collector voltage of Tr2 falls below the preset level on the base of Tr3. Transistor Tr5 and Tr6 form a high-gain, compound-emitter follower that isolates the output from C2.

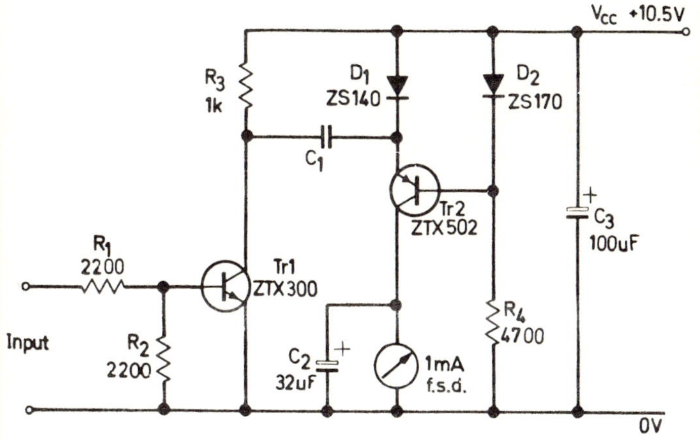

Frequency range	C1
10Hz--100Hz	1·0μF
100Hz–1kHz	0·1μF
1kHz–10kHz	0·01μF
10kHz–100kHz	1000pF
100kHz–1MHz	100pF

Fig. 7-9. A simple pulse counter that will operate over the range 10 Hz to 1 MHz by the choice of a suitable value for C1.

With the component values shown, the circuit will generate a staircase of 2 to 10 steps, with an input pulse rate of 50 Hz to 2 kHz. Figure 7-8 shows a 50-step staircase generated by replacing C1 with a 2200 pF capacitor. The input pulse rate in this case was 10 kHz.

The staircase generator can be used as a simple frequency divider if Tr5 and Tr6 are omitted and the output is taken from the collector of Tr4. In the interest of stability, division ratios of more than five are not recommended.

PULSE COUNTERS

A pulse counter provides an output voltage, or current, proportional to the input frequency. Figure 7-9 shows an example, and the similarity between this counter and the staircase generator will be evident. When Tr1 turns on, C1 charges via Tr1 and D1, and stores charge $Q = C1 \times V_{CC}$ (approximately).

When Tr1 turns off, C1 discharges through R3 and the emitter of Tr2; the charge of C1 is transferred to the collector circuit of Tr4, giving a mean current of $i = Q/t = Q \times f$, where f is the input frequency. That is, for i.e. full-scale deflection of a 1 mA meter at 100 kHz with $V_{CC} = 10V$, $i = Q \times f$. Therefore,

$$Q = \frac{i}{f} = \frac{10^{-3}}{10^5} = 10^{-8} \text{ coulombs. Hence,}$$

$$C1 = \frac{Q}{V_{CC}} = \frac{10^{-8}}{10} = 10^{-9} \text{ farads (1000 pF).}$$

In practice, a slightly higher value of C1 or V_{CC} is required because of the voltage drop across D1. Care must be taken that C1 can discharge completely when Tr1 is turned off—i.e., at 100 kHz, Tr1 will be turned off for 5 usec (square wave); therefore, C1 x R3 is less than 5 usec for C1 to discharge to 1 percent of V_{CC}. Resistor R3 is less than 1K.

The circuit shown in Fig. 7-9 will operate over the range of 10 Hz to 1 MHz by choice of C1 with a linearity of 2 percent of full scale on the lower ranges and 5 percent of full scale on the 100 kHz to 1 MHz range. It can be calibrated either by adjustment of the supply voltage or by an adjustable meter shunt.

Chapter 8

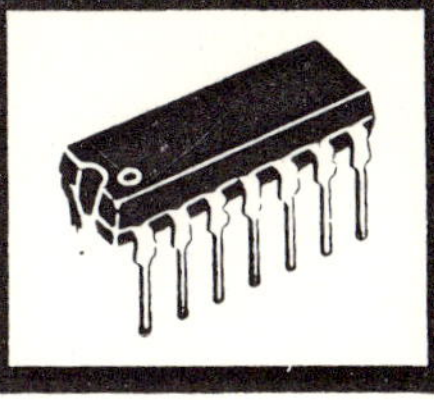

Schmitt Trigger Circuits

The Schmitt trigger circuit is a useful level-operated switch. Feedback via the common-emitter resistor causes the circuit to switch rapidly from on to off, and vice versa.

BASIC CIRCUIT

The circuit of Fig. 8-1A has two states. One state is with Tr1 and Tr2 off; the other with Tr1 off and Tr2 on. If V_{in} is increased from zero, when the transition voltage V_{T1} is reached, Tr1, which was nonconducting, now starts to conduct, cutting off Tr2, and the circuit rapidly changes state if the loop gain is greater than unity. If the loop gain is less than unity, the circuit will slowly change states and can be stopped with both transistors conducting. If V_{in} is now reduced, when the second transition voltage V_{T2} is reached, the current through Tr1 begins to reduce, and regenerative action causes Tr2 to come on, turning off Tr1.

It should be noted that V_{in} must be reduced to a value less than V_{T1} for the circuit to revert to its former state. The difference between V_{T1} and V_{T2} is referred to as the hysteresis or backlash of the circuit. Figure 8-1B shows the relationship between the collector current of Tr2 and V_{in2}.

It is difficult to predict the values of VT1 and VT2 accurately, but the approximate formulas for VT1 and VT2 given below may be of assistance to the circuit designer:

The upper threshold $V_{T1} =$

$$\frac{V_{BE1} + V_{cc}(R_E R_{L2} + R_E R_{L1}) - V_{BE\,(sat)}\, R_E R_{L2})}{R_{L1} R_E + R_E R_{L2} + R_{L2} R_{L1}}$$

The lower threshold $V_{T2} =$

$$\frac{V_{BE1} + (V_{cc} - V_{be2})\, R_E}{R_{L1} + R_E}$$

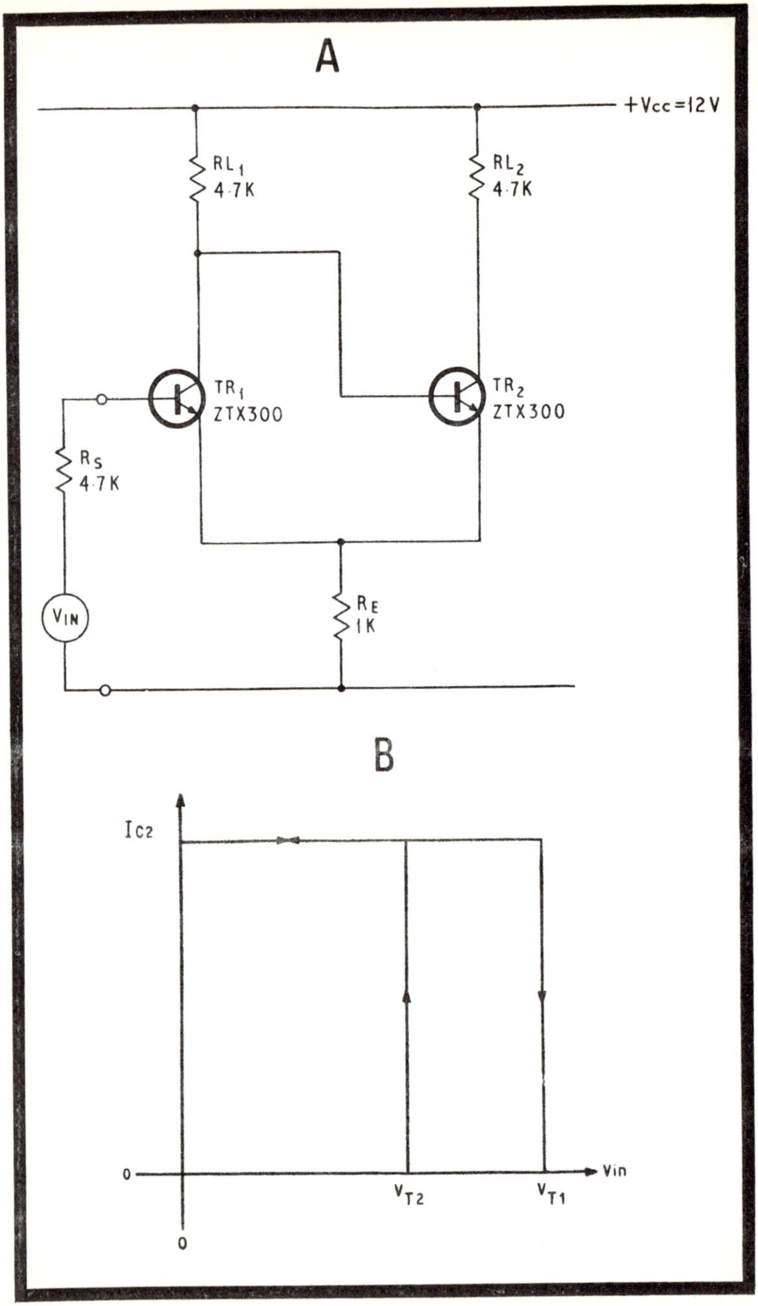

Fig. 8-1. Basic Schmitt trigger, and relationship between the collector current of Tr2 and V_{in}.

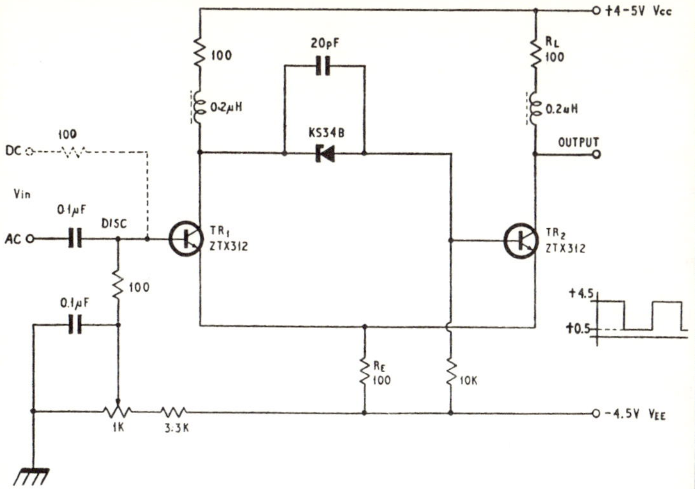

Upper trace—output waveform
at 50MHz (2V/cm).

Lower trace—Output waveform
at 10MHz (2V/cm).

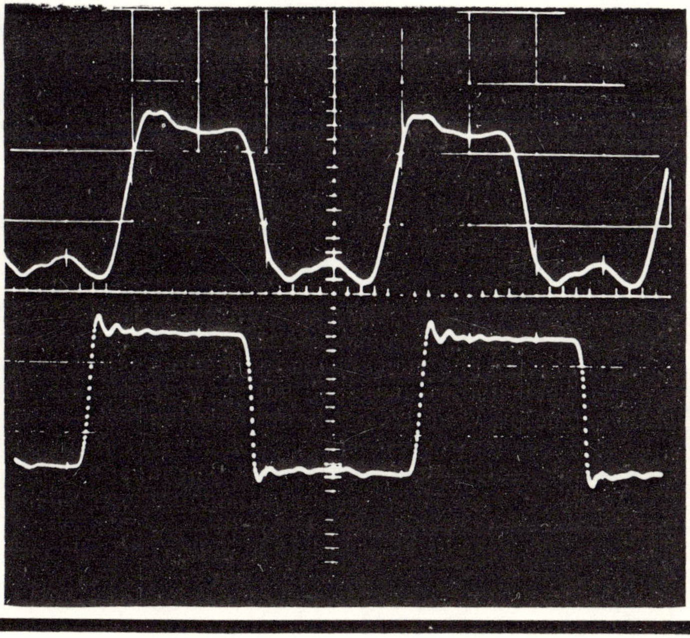

Fig. 8-2. High-speed Schmitt trigger, with waveforms.

Assuming V_{be} (sat) = 0.75V and $V_{be1} = V_{be2} = 0.65$V and that $R_s \ll R_n$

$$V_{T1} = 4.12\text{V}$$

$$V_{T2} = 2.63\text{V}$$

(V_{be1} and V_{be2} are the forward base-emitter voltage drops of Tr1 and Tr2, respectively, when their collector currents are equal.)

Measured values of V_{T1} and V_{T2} are 4.15V and 2.6V, respectively.

HIGH-SPEED SCHMITT TRIGGER

The circuit of Fig. 8-2 is a high-speed Schmitt trigger that operates from dc to 50 MHz. To eliminate storage delay times, the transistors are kept out of saturation by operating them in the current mode. Rise and fall times are less than 6 nsec.

The zener diode keeps the base of Tr2 at about 0V when Tr1 is on, the emitter current being given by

$$\frac{V_{EE} - V_{be}}{R_E}.$$

The values of V_{cc} and R_L are selected so that

$$\frac{V_{cc}}{R_L} \text{ is greater than } \frac{V_{EE} - V_{be}}{R_E}$$

and the transistor stays out of saturation.

The high-frequency chokes are included to compensate for capacitive loading of the Schmitt trigger, and may be omitted if the loading is light. Sensitivity is 1.5V at 50 MHz.

Chapter 9

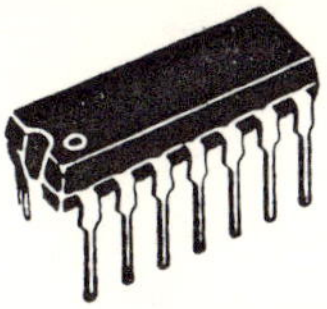

Complementary Circuits

The use of npn and pnp transistors in the same circuit can result in simplification of interconnections and a reduction in the number of resistive and capacitive components. The circuits in this chapter illustrate such applications.

TIMING RELAY CIRCUIT

In Fig. 9-1, with C1 completely discharged, charging current flows in the circuit R1-R2-R3-C1-R4. The voltage across R1 turns on Tr1, and the rising voltage on the base of Tr2 turns this transistor on. As there is a positive feedback loop, both transistors turn on and the capacitor charges, almost to the supply voltage, very rapidly. As the charging current falls off, a reverse process occurs, and both Tr1 and Tr2 turn off. The top side of C1 is tied to ground via R4, and the other side goes negative with respect to ground.

The capacitor discharges to the supply line via 5M resistor R3, giving a 5 sec delay. When the base voltage of Tr2 rises above the emitter voltage, the action is repeated. The reverse-biased base-emitter junction has a typical breakdown voltage of 7.5V, and this limits the negative potential on the base. (As 12K resistor R4 is in the breakdown circuit, this is permissible.) A positive pulse is applied every 5 sec to output transistor Tr3 as transistor Tr1 turns on. With the relay replaced by a 10K resistor, a pulse of 3 msec was obtained at the collector. The timing resistors conduct only during the charging part of the cycle; thus, the circuit consumes very little power.

DIRECT-COUPLED AMPLIFIER

The circuit in Fig. 9-2 uses transistors in a ring-of-3, and demonstrates the high gain that can be obtained with simple circuitry using directly coupled complementary types. The voltage gain is 3×10^4, or 90 dB. The high-frequency gain is determined by C3; with the value given, it is 3 dB at 20 kHz. Capacitor C3 promotes high-frequency stability.

The low-frequency gain is determined by C1, C2, and the source resistance, resulting in a hump before cutoff. With the values given, the voltage gain rises 4.5 dB at 40 Hz and falls to —3 dB at 25 kHz.

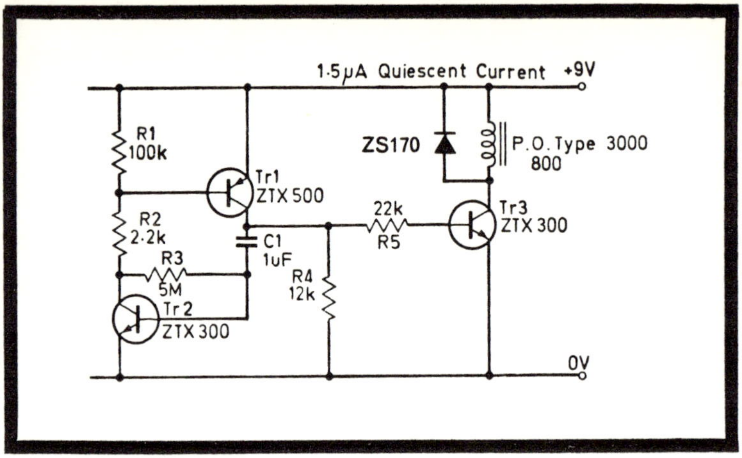

Fig. 9-1. Timing relay circuit.

150 mW AUDIO AMPLIFIER

This amplifier (see Fig. 9-3) is suitable for use as an intercom, using a second 35 ohm loudspeaker as a microphone. The two loudspeakers can then be interchanged by suitable switching. Design parameters and procedure follow.

For a power output of 0.15W into a 35 ohm load, an rms voltage of

$$E = \sqrt{PR} \text{ is needed;}$$

$$\text{Thus, } E = \sqrt{0.15 \times 35\ \ 2.3V\ rms}$$
$$= 2.3 \times 2\sqrt{2}\ p\text{-}p$$
$$= 6.5V\ p\text{-}p$$

Therefore, allowing for a minimum 1V collector-emitter potential, a 9V supply is needed.

Peak current needed from the output transistors is calculated from

$$I_m = \sqrt{(2P)/R}$$

$$= \sqrt{(2 \times 0.15)/35} = 93\ mA$$

Minimum h_{FE} of Tr3 and Tr4 at 100 mA collector current is 20. Thus, the maximum base current required is 5 mA and, to supply this, Tr2 is arranged to operate at 6 mA.

The ac collector load of Tr2 is

$$h_{FE} \times R \text{ (speaker)}$$
$$= 20 \times 35$$
$$= 700 \text{ ohms}$$

and the ac resistance

$$r_e \text{ of Tr2} = 27/I_c(\text{mA}) = 27/6$$

$$= 4.5 \text{ ohms}$$

Thus, the voltage gain of Tr2 is 700 ⁄ 4.5=150.

The input impedance of Tr2 is

$$r_e \times h_{FE}, \text{ in parallel with } \frac{R8}{\text{V gain, Tr1}}$$

$$= 4.5 \times 100, \text{ in parallel with}$$

$$\frac{68 \times 103}{150} = 225 \text{ ohms}$$

Thus, loss caused by R5 is:

$$225/1225 = 0.18$$

Therefore, voltage gain from the junction of VR1—R5 to the output is 0.18 x 150=27. Sensitivity at junction VR1-R5 for 0.15W output is 2330 ⁄ 27 = 85 mV.

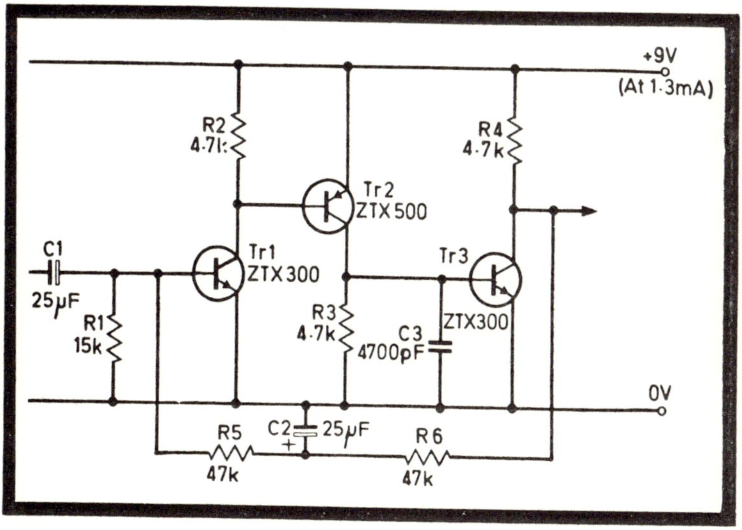

Fig. 9-2. Direct-coupled amplifier.

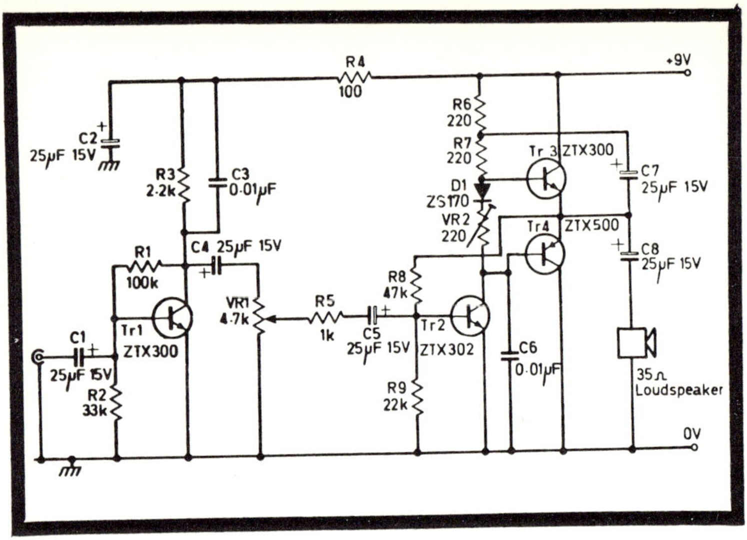

Fig. 9-3. 150 mW audio amplifier.

Transistor Tr1 is arranged to operate at 2 mA collector current:

$$r_e = \frac{27}{2} = 13.5 \text{ ohms}$$

The collector load is 2.2K; 4.7K and 1K, in parallel, =600 ohms. Therefore, the voltage gain of Tr1=

600/13.5=4.5. The sensitivity for full output 85/45=2 mV.

The input impedance is r_e x h_{FE} in parallel with 100K/V gain, Tr1. That is the input impedance is 13.5 x 100 ohms in parallel with 100,000/45 ohms.

Before the supply is applied, potentiometer VR2 should be set to zero, and subsequently adjusted to give 1 mA quiescent currents in Tr3 and Tr4.

The lower frequency limit can be extended to —3 dB at 40 Hz by increasing C8 to 250 uF.

The specifications for the design are:

Supply current with zero output	=9 mA
Supply current at 150 mW output	=40 mA
Power output	=150 mW at 10 percent total harmonic distortion
Sensitivity for 150 mW output	=2 mV rms
Input impedance	=850 ohms
Frequency response	= –3 dB at 200 Hz and 20 kHz

Chapter 10

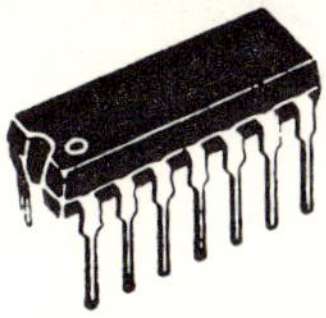

Diode Transistor Logic (DTL) Nor Gate

Figure 10-1 is an example of a DTL gate for low-voltage, low-power, noncritical applications. The gate is on when the input is up (3V to 5V or open circuit); it is off when the input is at 1V or lower (connected to the collector of a saturated transistor or to ground).

The noise immunity,(which is defined as the amplitude of noise on the ground required to turn on a gate held off by an on gate with worst-case Vout(on) = 0.3V attached to its input) is 0.7V at 55 deg C and 1V at 25 deg C.

Rise and fall times are 30 nsec and 200 nsec, respectively, for one gate load, and 50 nsec and 140 nsec for a 10-gate load (100 pF in parallel with 470 ohms, plus diode).

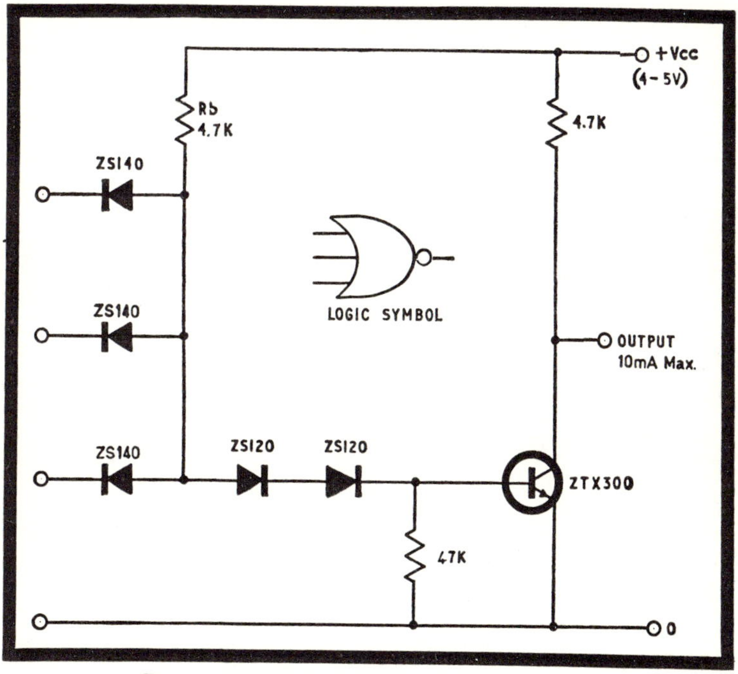

Fig. 10-1. Diode-transistor logic NOR gate.

The design procedure for the gate is outlined briefly below. The general requirement for low-power, low-speed logic should be kept in mind. Operating temperature range is —10 to 55 deg C.

The first requirement suggests a gate load of about 4.7K and a maximum collector current of 10 mA because the direct-current gain of the ZTX300 transistor at 10 mA and 25 deg C is 50. At −10 deg C, this is reduced by

$$50 \times \frac{0.7}{100} \times 35 = 12.25$$

to 37.75 (h_{FE} falls by 0.7 percent per degree centigrade drop in temperature).
The required base current is then

$$I_c / h_{FE} = 266 \text{ uA}.$$

For a bleeder resistor of 47K, tolerance 20 percent (max 56.7K, min 37.4K):

Max. bleeder current=

$$\frac{V_{be} \text{ (sat)}}{37.5 \times 10^3} = 20 \text{ uA}$$

Required current through R_b=266 + 20=286 uA; for V_{cc} min 4V and max sum of V_{be} + $2V_f$=2.4V. (V_f is the forward voltage drop of the series diode.)

$$R_b(\max) = \frac{V_{cc} - V_{BE} - 2V_f}{I_b + I \text{ bleeder}} = \frac{4 - 2.4 - 5.6K}{286}$$

Therefore, the **nominal** value of base resistor R_b is 4.7K.
Current through one worst-case input

$$\frac{V_{cc} - V_f \text{ (min)}}{R_b \text{ (min)}}$$

$$= \frac{4 - 0.6}{3.75 \times 10^3} = 0.9 \text{ mA}$$

$$\text{Fan out} = \frac{10}{0.9} = 11$$

SET-RESET BISTABLE

Cross coupling of two NOR gates results in the set-reset bistable, or logic flip-flop of Fig. 10-2. This circuit is useful for

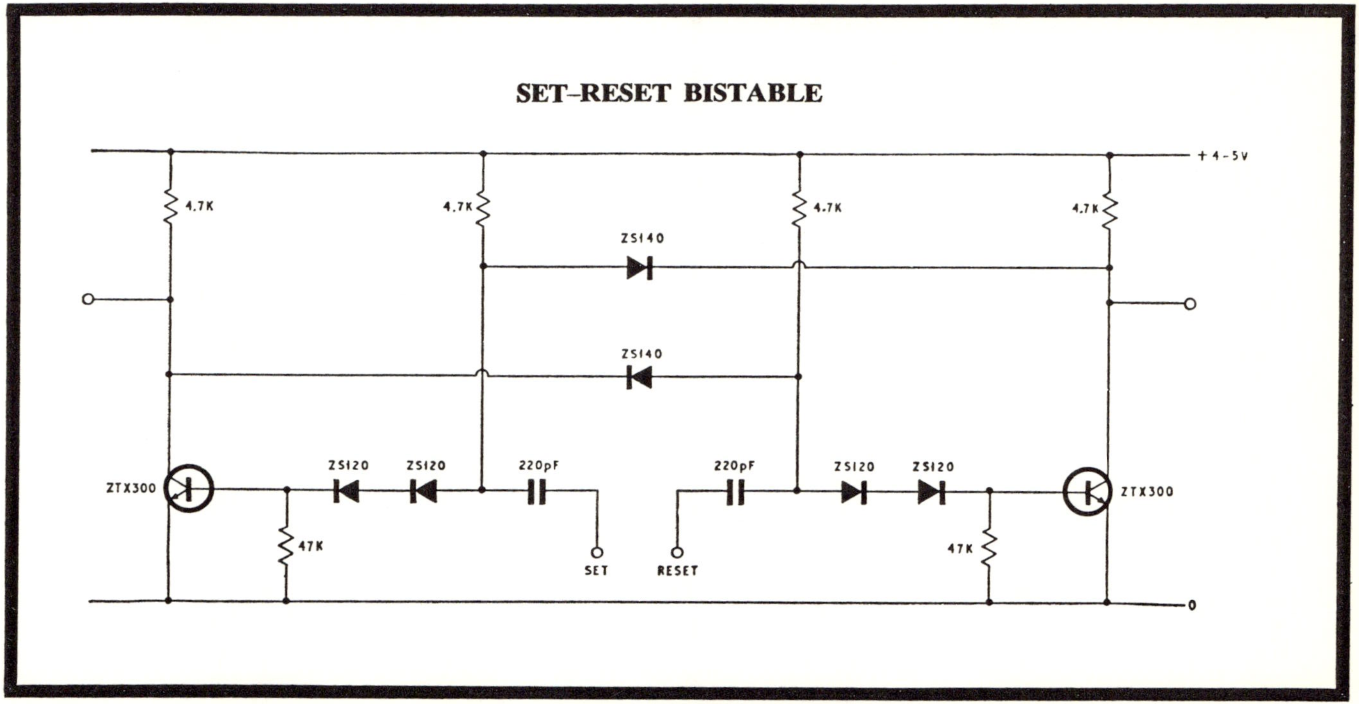

Fig. 10-2. Set-reset bistable multivibrator.

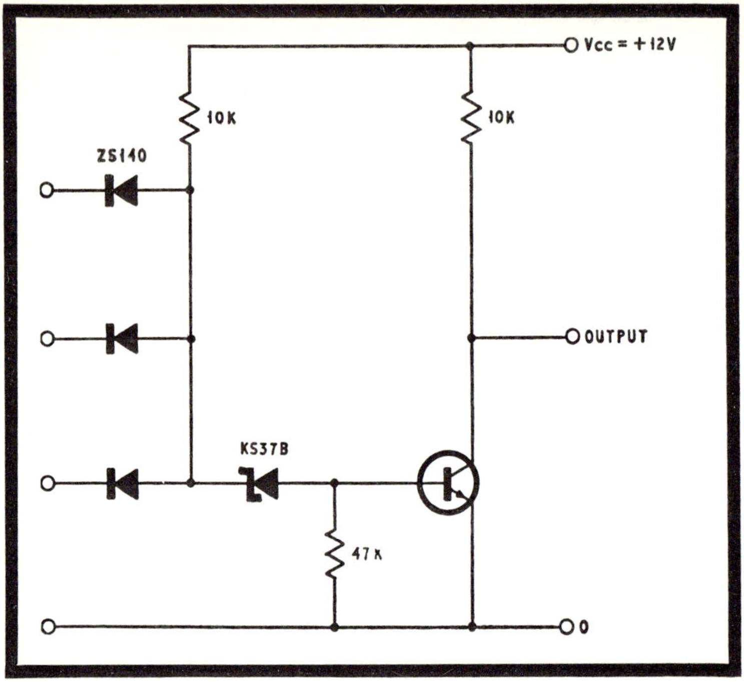

Fig. 10-3. High noise immunity gate.

control purposes where two separate signals are available to trigger the bistable between its two states.

Inputs are normally up and triggering occurs when the input is pulled down to ground (less than 1V) potential.

HIGH NOISE IMMUNITY GATE

Severe problems are encountered when conventional logic circuitry in control gear and instrumentation is placed in high noise environments; i.e., conditions in which switching transients are present in the line voltage supplies and in which stray electrostatic and electromagnetic fields may exist.

The high noise immunity gate of Fig. 10-3 has a typical dc noise immunity of 4.5V (4V min). This immunity will considerably reduce the possibility of spurious operation in low-speed systems. Fan out is 10 max. Temperature range is from minus 10 to 55 deg C.

Chapter 11

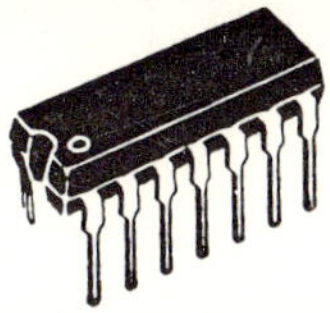

Avalanche Pulse Generator

When a resistor is connected between the base and emitter of a transistor, and voltage is applied between collector and emitter, a characteristic graph of the form shown in Fig. 11-1 is obtained.

The collector voltage can be increased up to a value of V_{CER}, which is dependent on the value of resistance. If the power is increased further, the collector-emitter voltage drops to a low value and there is a sharp increase in collector current. Transistors can easily be damaged or destroyed in this region, as the breakdown is caused by the collector current being channeled through a small area of the collector-

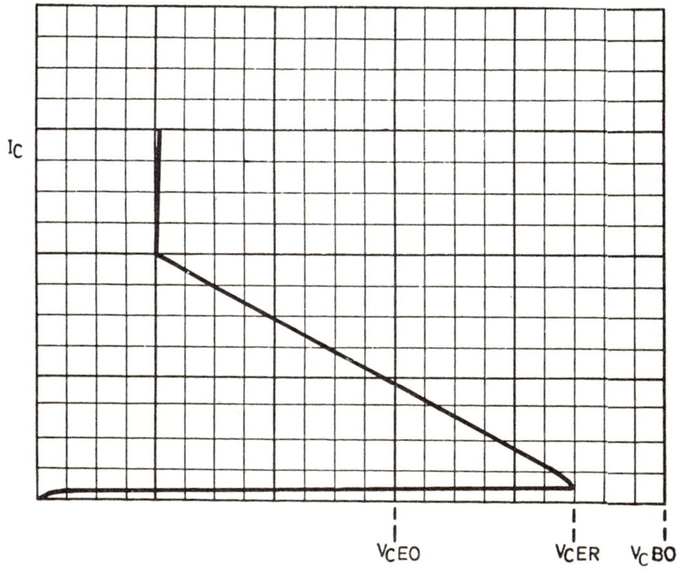

Fig. 11-1. Graph of I_C vs V_C.

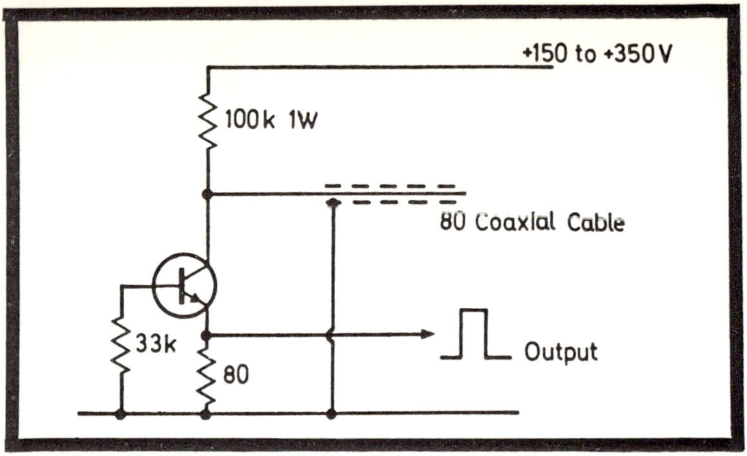

Fig. 11-2. Avalanche pulse generator.

base depletion layer. If the current is limited, however, use can be made of the very rapid transition into the breakdown region to produce short pulses with very fast rise times.

The circuit shown in Fig. 11-2 depends upon a charged delay line (the coaxial cable). The capacitance of the line is charged from the power supply until the voltage reaches the transistor V_{CER}, at which point the transistor breaks down. The delay line then discharges through the transistor and the matched load resistor, giving a pulse with very fast rise and fall times. Pulse duration is defined by length of the delay line. The charging rate, and hence the pulse repetition rate, is defined by the value of the supply voltage.

The pulse width is approximately 10 nsec per yard of 80 ohms coaxial cable. The rise and fall times are approximately 1 nsec. Output voltage of transistor ZTX302 (BCW14) is 30 to 60V. Repetition rate is up to 5 MHz. For transistor ZTX313 the output voltage is less than 20V for a 20 nsec pulse.

Chapter 12

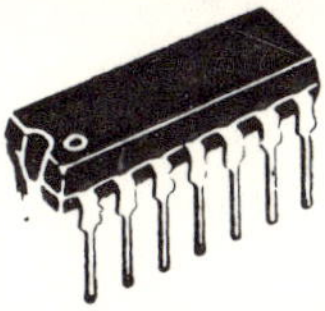

Regulated Power Supplies

Most electronic circuits require power supplies, and it is usually desirable for the power to be obtained from the ac lines. To minimize ripple and to provide a stable supply voltage, regardless of line and load variations, a regulated power supply is required. There are basically two types—series regulators and shunt regulators.

In the series regulator (Fig. 12-1), the regulator device is in series with the load. An error amplifier compares the output voltage with a reference source—usually a zener diode—and controls the current flowing in the regulator device in order to minimize the error. The power dissipated in the regulator is very low with no load, and rises to a maximum between half and full load. There is no inherent short-circuit protection, but this can be provided by additional circuitry, as shown in Fig. 12-2.

In the shunt regulator, the simplest form of which is a single zener diode as in Fig. 12-3, the regulator element is

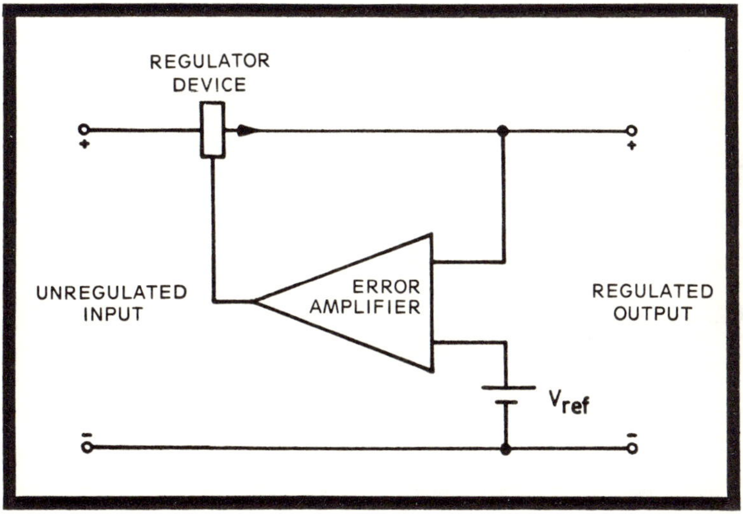

Fig. 12-1. Basic series regulator.

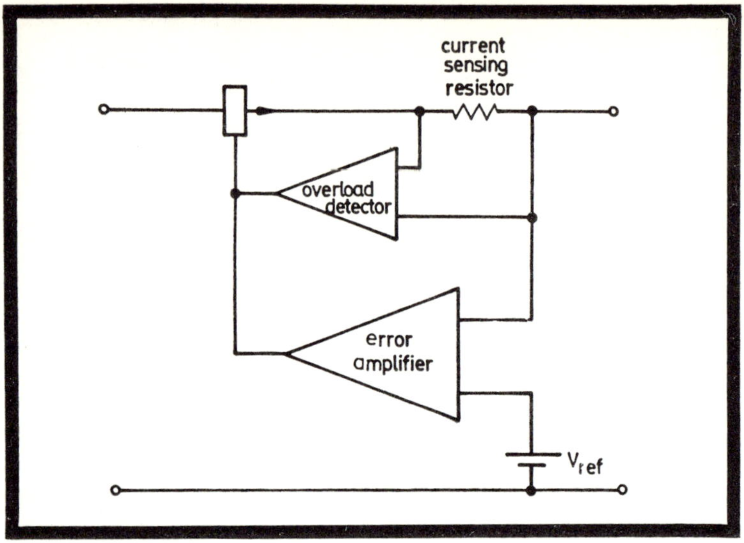

Fig. 12-2. Series regulator with current limiting.

placed in parallel with the load and is fed current by a series resistor. The error amplifier (Fig. 12-4) controls the current flowing through the regulator device to compensate for variations of load current and to maintain constant output voltage. The shunt regulator has inherent short-circuit protection, but its efficiency is low because of the power dissipated in the series resistor. For this reason, it is normally used only in low-current applications.

Figure 12-5 shows an extremely useful series regulator circuit that can be adapted to a wide range of output voltages and currents by choice of suitable components. The output

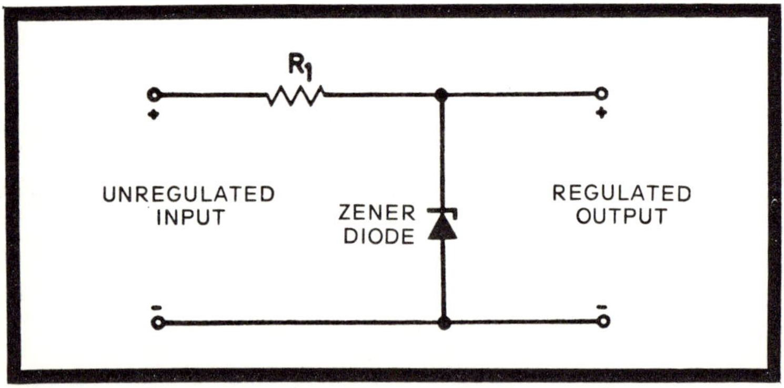

Fig. 12-3. Simplest form of shunt regulator.

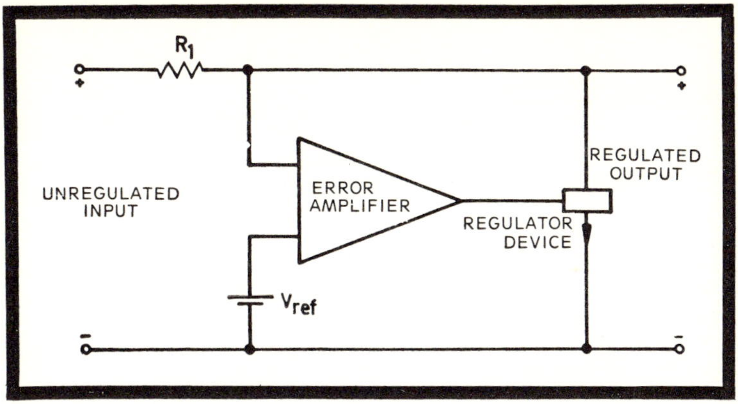

Fig. 12-4. Shunt regulator with error amplifier.

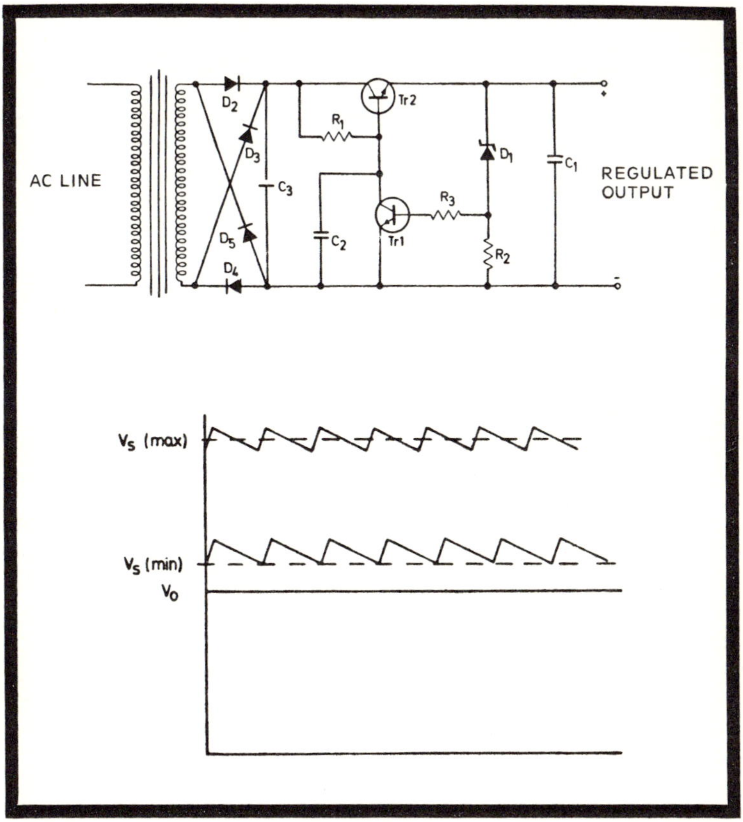

Fig. 12-5. A practical series-regulator circuit, with unregulated input conditions for high and low voltage.

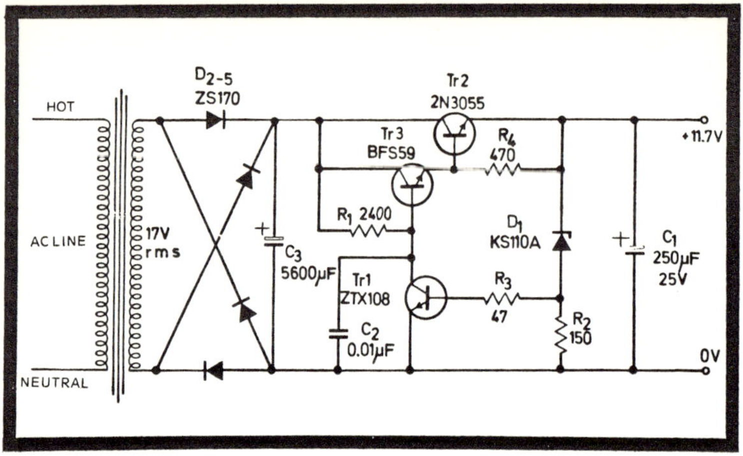

Fig. 12-6. Regulated 12V, 1A power supply.

voltage is fixed at 0.7V above the zener voltage of D1. In most cases, it will be necessary for Tr2 to be a Darlington pair of a low-power transistor and a high-power type because of the power dissipation. The approximate design procedure is described below, assuming 10 percent ac line variation.

DESIGN PROCEDURE, 12V-1A POWER SUPPLY

Choose a zener diode to give output voltage as near as possible to that required; i.e., for 12V nominal output, an 11V zener (KS110A) would give 11.7V. See Fig. 12-6.

Estimate the minimum input voltage as 10 percent or 2V higher (whichever is the greater) than the required output. This must be measured to the bottom of the ripple waveform, as shown in the waveforms of Fig. 12-5. For a 12V output, the minimum input voltage would be 14V.

Choose a suitable value of smoothing capacitor so that the ripple voltage is not more than about 10 percent of the output voltage; i.e., for a 1A output at 12V, the minimum ripple would be about 1.2V p-p. For a full wave rectifier, CV=It; therefore,

$$C = \frac{It}{V} = 8000 \text{ uF}.$$

In practice, the finite winding resistance of the power transformer tends to reduce the ripple, and a smaller value can be used. A 5000 to 6000 uF capacitor would be suitable.

It is rather difficult to calculate the output voltage required from the power transformer secondary because the regulation of the transformer varies considerably according to manufacturer,core material, and size.

In many cases, it is necessary to use a stock transformer; the voltage chosen should then be the nearest available **on the high side of the calculated value.**

To calculate the power dissipation in the transistors used, the maximum supply voltage must be known. This can be estimated from

$$Vs(max)=(V_{rms} + \sqrt{2})- \frac{V_r}{2} - V_d$$

This allows for 10 percent rise caused by line variations and a 10 percent fall caused by load regulation, the two factors canceling out. If we use a stock transformer with a 17V rms secondary, Vs (max) = (17 x 21 — 0.6 — 2.0 = 21.4V (average).

With no load, the supply voltage will be approximately 10 percent higher than this (i.e., 24V), and the voltage rating of filter capacitor C1 would exceed this by at least 20 percent.

The minimum supply voltage with a 17V transformer (Fig. 12-6) will be $Vs(min)=17 \times \sqrt{2} \times 0.8) - 1.2 - 2.0 = 16V$

The power dissipation in the regulator device, Tr2, will be $I_L \times (V_S \max - V_O)$, where I_L is the maximum load current and V_O is the output voltage. The power calculates as 1 (21.4 — 11.7) = 9.7W.

The current and power required necessitate a Darlington pair. For the power transistor, a 2N3055 can easily handle the required current, voltage, and power. To maintain a safe junction temperature at ambient temperatures up to 50 deg C, the thermal resistance of the heat sink should not exceed 10 deg C per watt.

If we assume a minimum current gain of 30 at 1A, the base current of the power transistor will be 33 mA and the maximum power dissipation in the driver transistor (Tr3 in Fig. 12-6) will be 9.7/30 = 323 mW. This is within the capabilities of a BFS59 for ambient temperatures up to 75 deg C.

The maximum base current I_b required by the BFS59 will be $33 / h_{FE\,min}$=0.83 mA. Resistor R1 must be able to supply this current at the minimum supply voltage, V_s (min), of 16V:

$$R1(max) = \frac{Vs(min) - V - V_{be2} - V_{be3}}{I_b}$$

Assuming V_{be2}=1.0V and V_{be3}=0.7V,

$$R1(max) = \frac{16 - 11.7 - 1.0 - 0.7}{0.83\ mA}$$

$$= 3.13K$$

If we allow for a 20 percent end-of-life tolerance for R1, its nominal value will be 2.6K. The nearest preferred value is 2.4K.

With R1 at the **low end** of its tolerance, its value will be 1.92K, and, with the maximum (no load) supply voltage of 24V, the current will be 5.5 mA. Hence, the power dissipation in Tr1 will be 5.5 x 13.4 = 74 mW. This is well within the ratings of a ZTX108. Tif the current in R1 were found to be excessive with maximum supply voltage, it would be necessary to redesign for a higher transformer output or to use higher-gain transistors.

Resistor R3 is chosen to limit the base current to Tr1 to a safe value under fault conditions. Its value should be as small as possible, and a suitable value is 47 ohms. Resistor R4 is chosen to shunt away the collector-base leakage current of the 2N3055. A value of 470 ohms is suitable; the extra current required from Tr3 is negligible (2 mA).

Capacitor C1 reduces the output impedance of the power supply at high frequencies (above 10 kHz); 250 uF is a suitable value. Capacitor C2 is chosen to suppress high-frequency oscillation. Its minimum value should be determined by test; 0.01 uF will usually be sufficient.

When a current limit is required, this can be provided as shown in Fig. 12-7. When the V_{be} of Tr4 exceeds 0.7V, transistor Tr4 starts to conduct and diverts base current from Tr3, hence limiting the output current to a preset value. See Fig. 12-8. The component values required can be calculated as described below.

Choose R5 to drop about 1V at maximum output current. For 1A output, R5 = 1K.

Recalculate R1, allowing for the additional 1V drop across R5. This will also increase the minimum supply voltage V_s(min) by 1V. Using a 17V rms transformer as before:

$$V_s(\text{min} = 16\text{V}$$

$$R1(\text{max}) = \frac{16 - V_o - V_{be2} - V_{be3} - I}{I_b}$$

$$= 1.93\text{K}$$

Hence, the maximum nominal value of R1 (assuming 20 percent ultimate tolerance) is 1.6K.

Resistor R7 is chosen to limit the base current of Tr4 to a safe value under fault conditions; 47 ohms is suitable.

The transistor operates under low-voltage, low-current conditions; type ZTX108 is suitable.

VARIABLE POWER SUPPLY

This circuit gives an output voltage (0-30V) that can be adjusted between zero and 30V, at currents up to 1A. It has current-limiting overload protection. See Fig. 12-9.

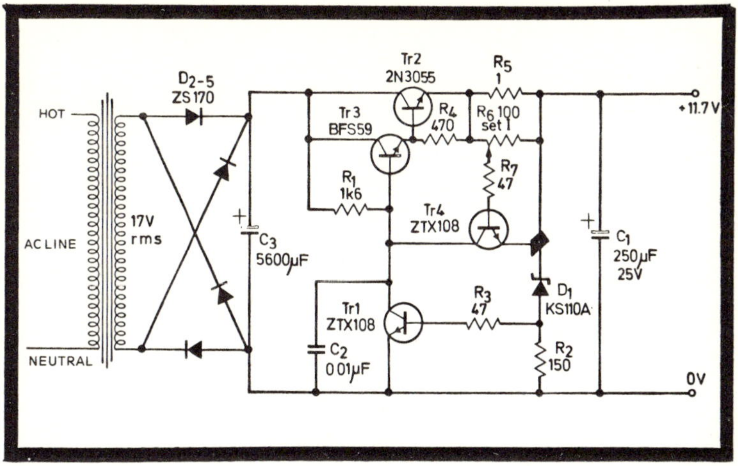

Fig. 12-7. Another regulated 12V, 1A power supply, with current limiting.

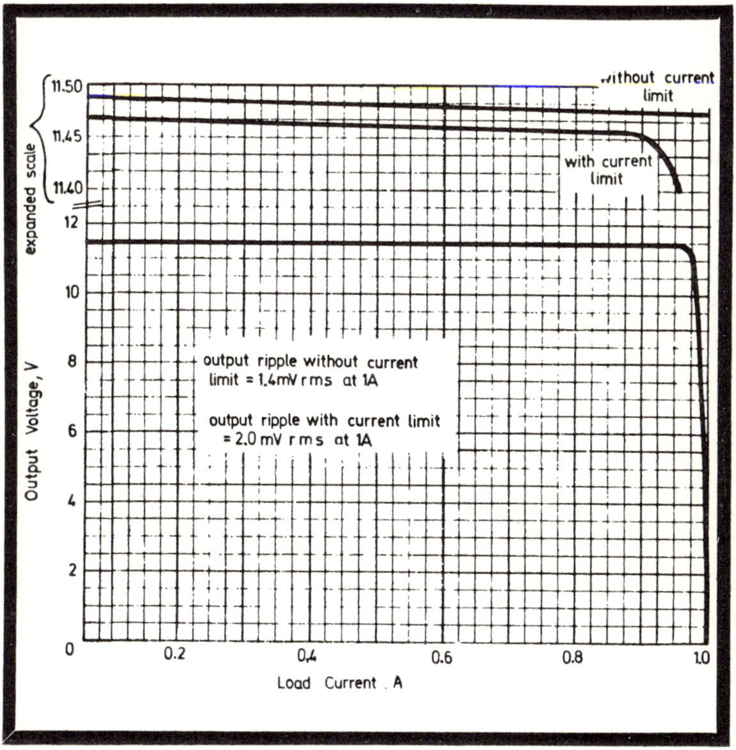

Fig. 12-8. Output voltage / load current graph.

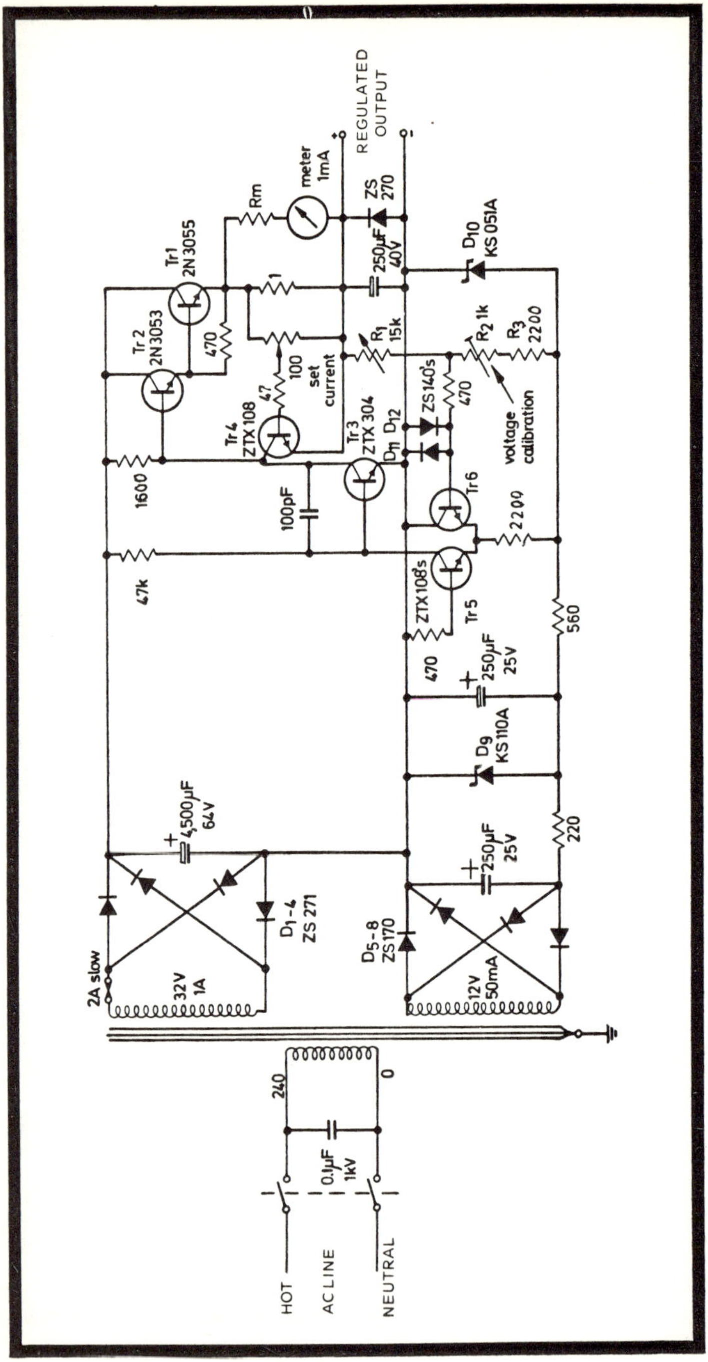

Fig. 12-9. A 0 to 30V, 1A variable power supply.

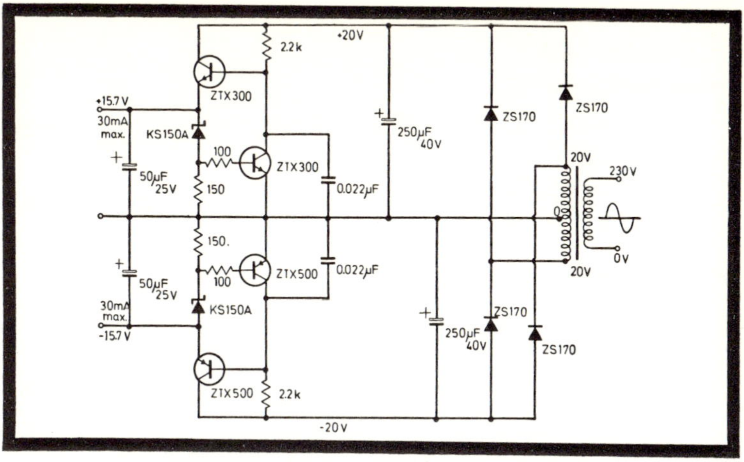

Fig. 12-10. Power supply for operational amplifier.

Transistors Tr5 and Tr6 form the error amplifier. The voltage on Tr6 base is compared with that on Tr5 (zero) and the difference is further amplified by Tr3, which controls the regulator device, Tr1-2. Because the voltage at point A must always remain close to ground, the current flowing through R2 and R3 must remain constant. If the base current of Tr6 is neglected, this same current flows through R1 so that the output voltage Vo Vr x R1—R2 + R3; i.s., the output voltage is directly proportional to R1, which can be a wirewound potentiometer or a bank of fixed resistors selected by switches.

The reference voltage is provided by zener diode D10. The SKO51A was chosen for its low temperature coefficient. Transistor Tr4 provides overload protection in the same way as previously described.

The dc output resistance is approximately 10 milliohms and there is 0.1 percent change in output voltage for a 10 percent increase or decrease in line voltage. Ripple at 10V output and 1A load is 1.5 mV rms.

Note: Transistor Tr1 should be mounted on a heat sink of not more than 2 deg C per watt thermal resistance. Transistor Tr2 should be fitted with a heat-dissipating clip; e.g., Redpoint 5F.

Resistor Rm is chosen to give a full-scale deflection of 1V on the output current meter; Rm is 900 ohms if the meter resistance is 100 ohms.

OPERATION AMPLIFIER POWER SUPPLY

Figure 12-10 shows a simple plus and minus 15V power supply that will give up to 30 mA output for as many as six ZLD709 operational amplifiers (not shown). It is basically the same circuit as the 12V power supply described above.

Chapter 13

Miscellaneous Circuits

This chapter describes the design and specification parameters for generally useful circuits for test equipment, etc., with some discussion of zener breakdown in transistors used as diodes, temperature drift, and similar subjects.

AC-MILLIVOLTMETER RECTIFIER BRIDGE

This circuit will give full-scale deflection on a 100 uA linear-scale meter for an input of approximately 20 mV rms over a bandwidth of 800 kHz.

The type of rectifier bridge shown in Fig. 13-1 requires an input of 220 uA rms (310 uA peak) for full-scale deflection of a 100 uA meter. Transistor Tr2, therefore, is biased at a collector current of 0.5 mA to ensure that adequate output current is available.

Resistor R5 is chosen to drop slightly less than half the supply voltage (to allow for the potential drop across D1). A suitable value is 8.2K.

Resistor R4 is chosen to drop approximately 0.8V; therefore, R4 is 1.5K.

The collector current for Tr1 is a compromise between input impedance and open-loop gain (and hence linearity). There is, however, no advantage in using a collector current greater than about 0.5 mA because the ac resistance, r_e, of the transistor is then falling below the external emitter resistance R1 (100 ohms). A bias of 340 uA is used in this design.

Resistor R3 is chosen to drop about 0.1V. Since Tr1 requires approximately 10 uA base current, R3 equals 10K.

A flat response down to about 10 Hz is usually required; therefore, large-value coupling and decoupling capacitors are necessary. The values shown in the circuit have been found to be adequate.

The performance of the circuit was as follows:

Sensitivity: 22 mV for full-scale deflection. (Can be adjusted by variation of R1.)

Bandwidth: 10 Hz to 800 kHz (plus 0, minus 5 percent)

Input impedance: 10K

No nonlinearity was detectable by comparison with Solartron VM1484 true rms voltmeter.

MILLIVOLTMETER PREAMPLIFIER

This circuit will increase the sensitivity of the rectifier stage already described to 1 mV full-scale. (See Fig. 13-2.)

To achieve maximum voltage gain, Tr2 load resistance should be as high as possible. Consequently, Tr2 is operated at a low collector current (equal to approximately 300 uA) and a 22K dc load is used.

The base current for Tr2 is derived from R6 and R7; the average current gain of ZTX302 approximates 100 at 300 uA. Therefore, base current equals 3.0 uA. Current through R5 approximates 1.3 uA. Hence, for a collector-base voltage of 2.0V (V_{ce}=2.6V), R6=2/4.3 uA=470K. Resistor R5 reduces the effect of transistor gain spread on the bias and for a large overload, reduces the recovery time of the stage.

For minimum noise, Tr1 is operated at a very low collector current (approximately 100 uA). It is biased in the same way as is Tr2.

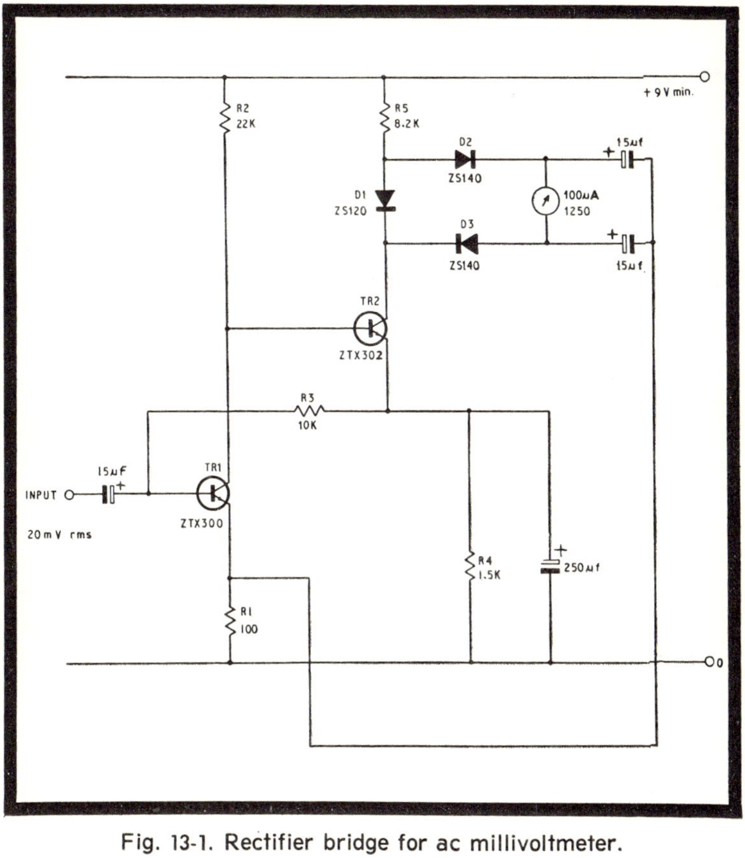

Fig. 13-1. Rectifier bridge for ac millivoltmeter.

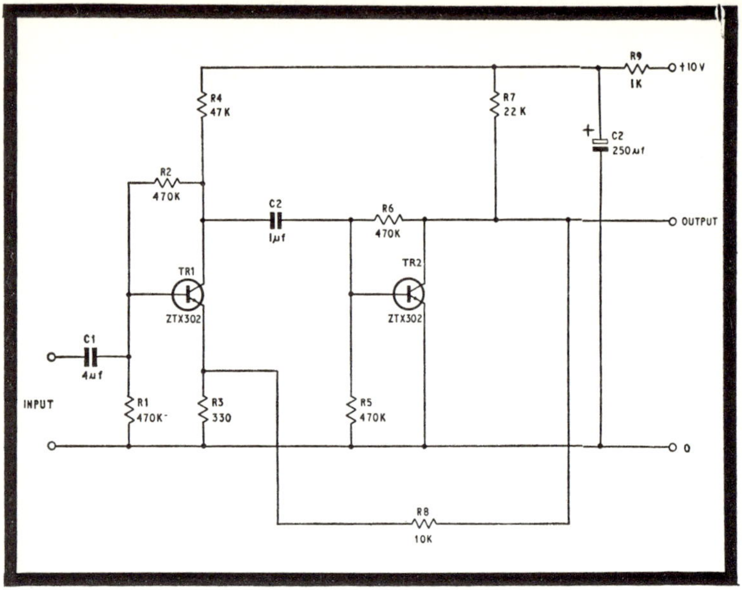

Fig. 13-2. Millivoltmeter preamplifier.

The input impedance is approximately 40K without feedback. When the feedback is connected, the input impedance becomes greater than 100K. Resistor R3 is adjusted to give the voltage gain required. For a gain of 22, the value of R3 is approximately 330 ohms. To avoid the possibility of low-frequency oscillation with a battery supply, the preamplifier is heavily decoupled by R9 (1K) and C3 (250 uF).

Performance was as follows:

Bandwidth (used with rectifier stage) equals 800 Hz (plus 0, minus 5 percent).
Noise with 1K source is less than 5 uV.
Voltage gain equals 22 (adjustable).
Input impedance is 100K.

WIDE-RANGE WEIN BRIDGE OSCILLATOR

The oscillator shown in Fig. 13-3 will operate over a wide frequency range, from less than 10 Hz to 5 MHz. The output is 1V rms with less than 0.2 percent total harmonic distortion (at 1 kHz). The circuit shown has five ranges:

(1) 15 Hz to 200 Hz
(2) 150 Hz to 2 kHz
(3) 1.5 kHz to 20 kHz
(4) 15 kHz to 200 kHz
(5) 150 kHz to 2 MHz

The frequency of oscillation is given by $f_0 = 1/(2\pi RC)$.

VERY LOW DISTORTION WEIN BRIDGE OSCILLATOR

This circuit (Fig. 13-4) gives a very low distortion output over a restricted frequency range. It is primarily intended to enable distortion measurements of audio amplifiers to be made without the inconvenience of tuned filters.

Design

At 1V rms output (normal level for the R53 thermistor), the worst-case load on Tr3 is approximately 220 ohms (R11 in parallel with 4.7K and 500 ohms plus thermistor and Wein network). Therefore, peak alternating current in Tr3=

$$\frac{\sqrt{2}}{200} = 7.07\,\text{mA}$$

so Tr3 is biased at 10 mA to provide adequate drive under all conditions.

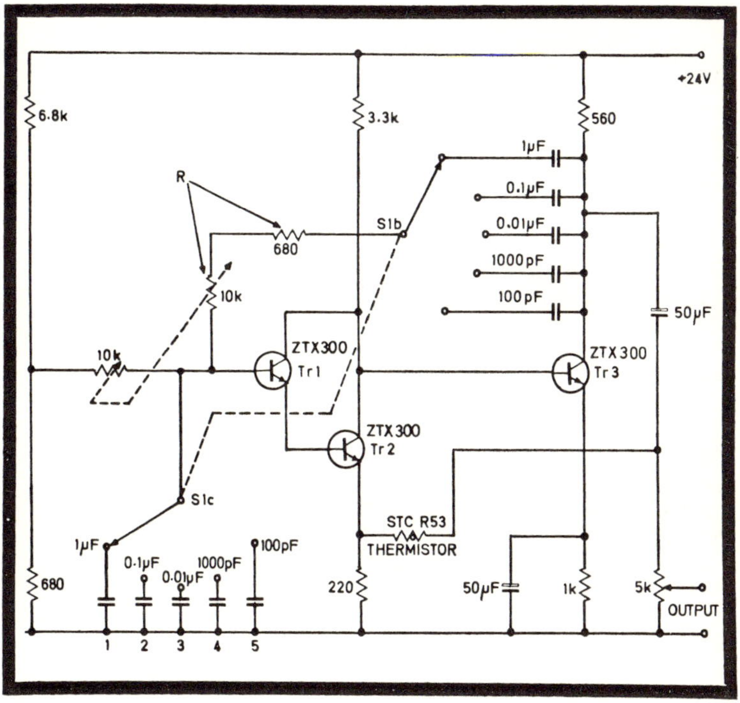

Fig. 13-3. Wide-range Wein bridge oscillator.

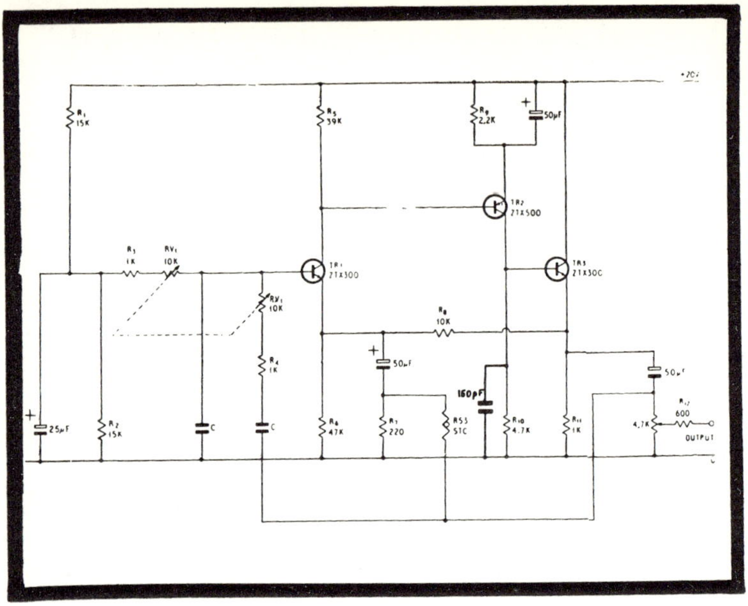

Fig. 13-4. Very low distortion Wein bridge oscillator.

Resistor R11 is chosen to drop half of the supply voltage; 1K is suitable. Resistor R10 is chosen as five times R11. The nearest standard value is 4.7K. Resistor R9 is chosen as half of R10; use 2.2K.

The collector current of Tr2 is chosen to drop half of the supply volts plus 0.7V (V_{be} of Tr3) across R10:

$$I_{c2} = \left(\frac{10.7}{4.7} + 0.1\right) \text{ mA}$$

(where 0.1 is the base current of Tr3)

$= 2.4$ mA

Therefore, the drop across R9 is approximately 2.4 x 2.2=5.3V, and the maximum base current required by Tr2 is approximately 55 uA.

For lowest noise and to achieve a high input impedance, Tr1 is operated at a low collector current (213 uA).

$$\text{Thus, R5} = \frac{\text{Voltage drop across R9} + V_{be} \text{ of Tr2}}{I_c \text{ of Tr1} - I_b \text{ of Tr2}}$$

$$= \frac{5.3 + 0.7}{213 - 55} \times 10^6$$

$$= 39\text{K}$$

Resistor R6 has half of the supply volts dropped across it, and hence its value is 10⁄0.213K—approximately 47K, which is used.

Resistors R1 and R2 are chosen to bias Tr1 at half of the supply voltage. Their exact value is unimportant; values of 10K to 22K are suitable. The exact value of R8 is also unimportant, but it should be large compared with R11 and small compared with R6. A value of 10K is suitable. To give the correct operating conditions for the thermistor, R7 was made 22 ohms. Ganged potentiometer RV1 is chosen so that its maximum value is much less than the input impedance of Tr1 (about 1M). A value of 10K for each leg is suitable.

Resistors R3-R4 are chosen to give the frequency ratio covered on each range—1K gives a max-to-min ratio of 11 to 1, providing a small overlap on decade ranges. If a more open frequency scale is required, increase R3-R4 to 3.3K. This gives 4 to 1 coverage on each range.

Performance

For the circuit values shown in Fig. 13-4, the frequency range was greater than from the expected 10 Hz to 100 kHz. Harmonic distortion, calculated to be less than 0.01 percent at 1 kHz, actually measured 0.004 percent. The output is approximately 1.0V rms. Output impedance is about 600 ohms.

PROCESS TIMER

This circuit will hold a relay on for a period between 1 and 10 sec. The section is initiated by a negative input pulse, which may be derived from other circuitry or from a simple switch (see Fig. 13-5).

The circuit was designed to operated a 12V, 400 ohms relay. If the relay coil resistance has a tolerance of 20 percent, the maximum current required equals 12⁄400 x 0.8 = 37.5 mA. At this current, the current gain of a ZTX300 is 86 percent of the 10 mA value. Also, to ensure satisfactory operation at temperatures down to minus 10 deg C, a further 25 percent fall in gain must be allowed for. Therefore, the worst-case current gain of Tr3 = 50 x 0.86 x 0.75 = 32.2.

The maximum base current required by Tr3 equals 37.5 mA⁄32.2 = 116 mA. The **maximum** value of R6 is given by

$$\frac{V_{cc} - V_{be}\,(Tr3)}{I_b\,(Tr3)} = \frac{12 - 0.7}{1.16} = 9.74K$$

Allowing a 20 percent life tolerance for R6, the nominal value is 9.75K⁄1.2 = 8.12K. We used 8.2K.

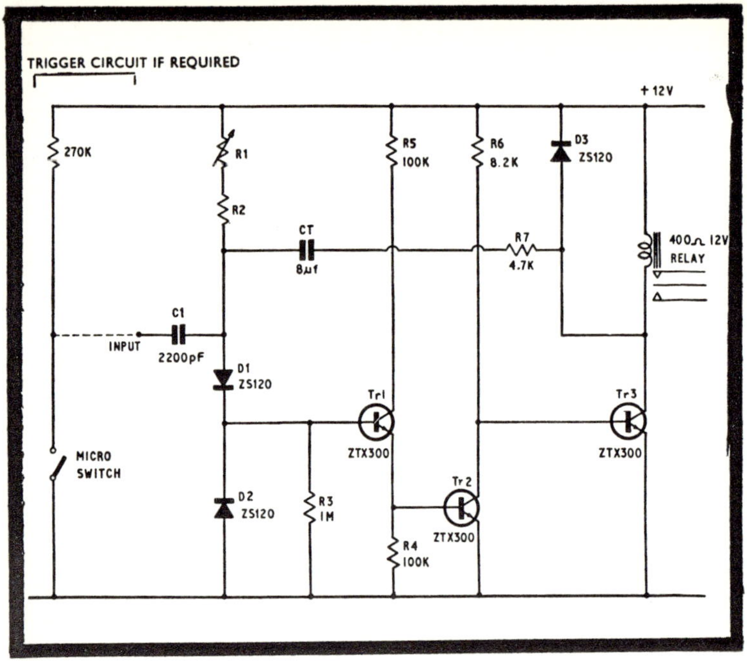

Fig. 13-5. Process timer.

The **minimum** (20 percent low) value of R6 = 8.2K x 0.8 6.5K. Therefore, the maximum **current** drawn from Tr2 ≥ V_{CC}—R6(min) = 12 /6.56 mA = 1.83 mA. At this collector current, the current gain of ZTX300 is 80 percent of the 10 mA value, and operation at minus 10 deg C results in a further 25 percent reduction: Tr2 h_{FE} (min)=50 x 0.8 x 0.75=30.

The base current required by Tr2 under worst conditions is: I_b (TR2)=1.83 mA/30=61 uA.

The bleeder resistor, R4, at the base Tr2 takes a nominal current of 7 uA, which is five times greater than the collector-base leakage current (I_{cbo}) of Tr2 at 55 deg C. If R4 is 20 percent low, this current increases to 9 uA. Therefore, the maximum current that Tr1 must supply equals 61 + 9 = 70 uA.

At 70 uA, the current gain of Tr1 (ZTX300) is 43 percent of the 10 mA value, and minus 10 deg C operation reduces this a further 25 percent. Therefore, h_{FE} (Tr1)=50 x 0.43 x 0.75= 16.1. The base current required by Tr1 equals 70 uA/16.1, or approximately 4.3 uA. The drop across D1 and the bases of Tr1 and Tr2 is about 2.1V.

Under quiescent conditions, the anode of D1 rests at approximately 2.0V and the collector of Tr3 at 12V. When the circuit is triggered, Tr3 collector voltage falls to approximately zero. This change is transmitted through timing capacitor C_t, driving D1 anode to minus 10V. Now C_t starts to charge from minus 10V toward 12V. However, when C_t has charged to 2V, D1 and Tr1 and Tr2 start to conduct, and the circuit reverts to its quiescent state. The monostable period is found as follows:

$$[2 - (-10)] = [12 - (-10)] \times [1 - e^{-\frac{t}{RC}}]$$

$$\frac{12}{22} = 1 - e^{-\frac{t}{RC}}$$

$$\text{therefore, } e^{-\frac{t}{RC}} = \frac{10}{22}$$

$$\text{and } \frac{t}{RC} = 0.787$$

It can therefore be seen that the monostable period is given by multiplying the time constant C_t x (R1 and R2) by 0.787. For a timing range of 1 to 10 sec, a timing capacitor of 8 uF is convenient, and gives

R2 = 200K
R1 = 2M pot

The purpose of diodes D1-D2 is to protect Tr1 and Tr2 from the 12V **negative transient** when Tr3 turns on. Diode D3 absorbs the inductive spike produced by the relay coil when Tr3 turns off. Resistor R7 was found necessary to prevent the circuit from free-running caused by ringing of the relay coil. Resistor R5 (100K) limits the current flowing in the bases of Tr1 and Tr2 to 100 uA.

USING THE BASE-EMITTER JUNCTION OF ZTX300 AS ZENER DIODE

Histograms of breakdown voltages of reverse-biased base-emitter junctions with two values of reverse currents are shown in Figs. 13-6 and 13-7. The typical breakdown voltage is 8V, and the devices exhibit a hard characteristic.

Slope impedance at 20 mA is typically 20 ohms, and maximum dissipation is 200 mW. The temperature coefficient is +5.45 mV per deg C or +0.059 percent per deg C.

A sample batch of ZTX310 transistors had typical breakdown voltages in the 6 to 7.5V range, with similar slope resistance to the ZTX300.

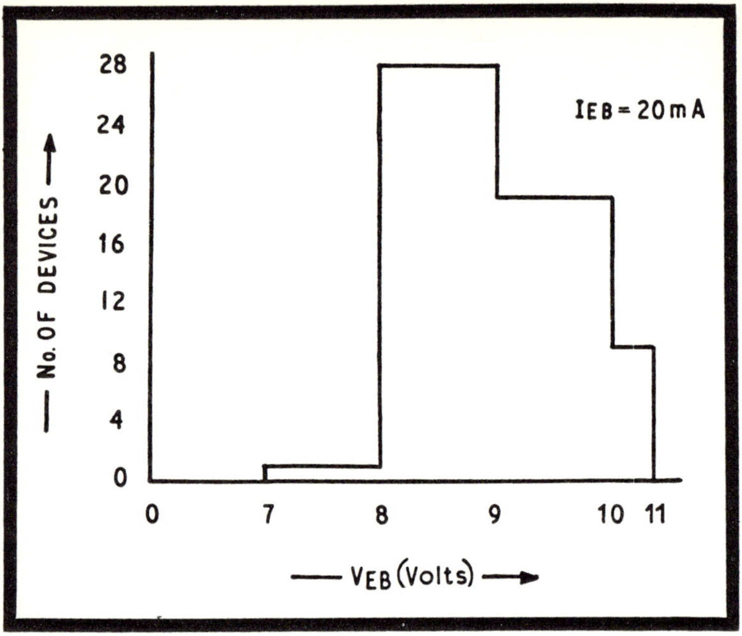

Fig. 13-6. Histogram of V_{EB} at 20 mA.

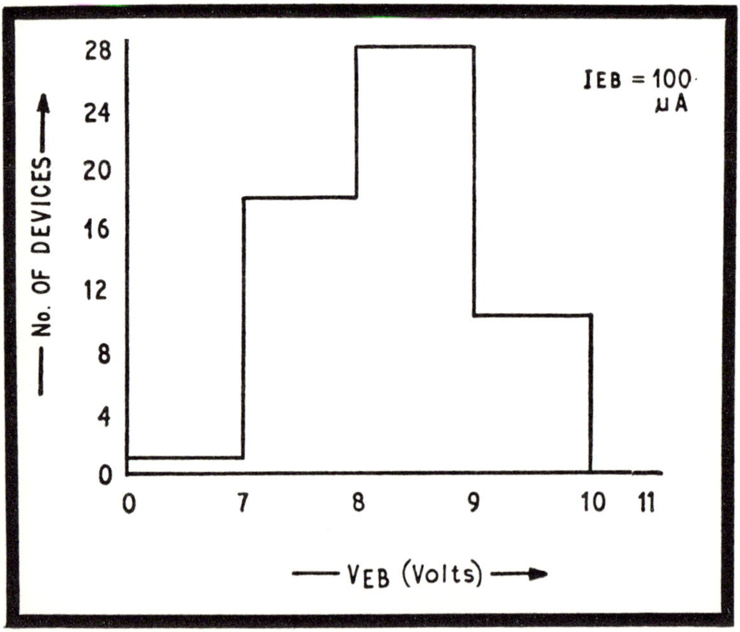

Fig. 13-7. Histogram of V_{EB} at 100 uA.

Degradation of Gain During Breakdown of Junction

The emitter-base junction of an E-line transistor may be used satisfactorily as a zener diode, as stated above. If, however, the device is required to act as both a zener diode and a transistor, a **marked decrease** in direct-current gain may result.

The degradation of current gain is dependent upon the amplitude and duration of the avalanche current and the surface condition of the particular transistor. It is attributed to high-energy electrons, liberated by avalanche multiplication, colliding with silicon atoms at the silicon-to-silicon-dioxide interface. The effect is to produce additional surface states, within or near the depletion layer, causing a reduction in current gain. The degree of degradation is difficult to predict, as it depends upon the avalanche current, the ionic charge in the oxide (which is dependent upon processing), the surface state of the transistor (also process dependent), and the transistor geometry.

Using the ZTX300 as a Diode

Three possible methods of connection are described:

1. Collector-base (with emitter connected to base)
This connection gives a diode with a medium recovery time, small capacitance, and high reverse voltage determined by the collector-base junction.
2. Base-emitter (collector connected to base)
The diode has a short recovery time, small capacitance, and a low reverse voltage determined by the base-emitter junction.
3. Base-emitter (collector connected to emitter)
The diode has a long reverse recovery time, a large capacitance, and a low reverse voltage, determined by the base-emitter junction.

Typical Recovery Time and Capacitance

The recovery time quoted is the interval between the negative input pulse attaining 10 percent of its final value and the reverse current attaining 10 percent of its maximum value. See Fig. 13-8, and Fig. 13-9, which graphs the variation of capacitance with reverse voltage.

Junction Temperature Measurements of ZTX300

The maximum power dissipation of the ZTX300 in free air at 25 deg is 300 mW.

The thermal resistance, junction to ambient, 0JA=0.33 deg C per mW. Mounted on a Redpoint heat sink, type 92-DC, the thermal resistance is reduced to 0JA = 0.21 deg C per mW, enabling 480 mW to be dissipated at 25 deg C ambient.

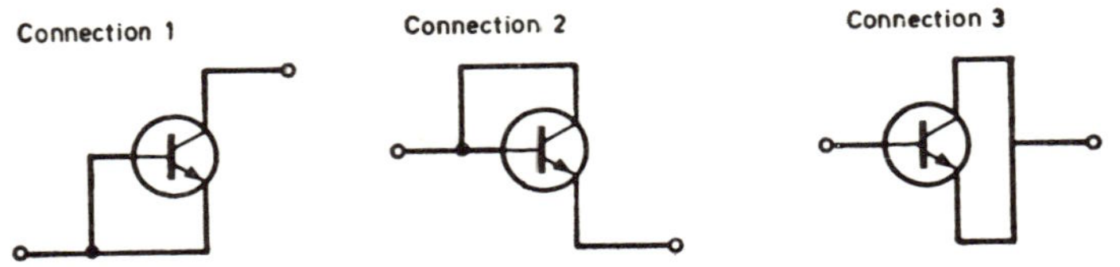

Method of Connection	Recovery Time nsec $I_F = I_R = 10mA$ $R = 50\Omega$	Capacitance pF		
		at zero voltage	at reverse voltage	
			−10V	−5V
1	89	7	3	
2	2.7	7		3.5
3	244	13		5.5

Fig. 13-8. Typical recovery time and capacitance.

With two transistors mounted in the same Redpoint heat sink, the heat sink should be clamped to a small aluminum plate to enable a worthwhile increase in power dissipation to be achieved. When clamped to an aluminum plate 1.5 in. x 1 in. thick, x 18-gage, 480 mW can be dissipated in **each** transistor.

If the transistor is clamped directly to a 1 in. x 1 in. x 18-gage thick aluminum plate, a further reduction in thermal resistance occurs, so that Θ_{JA}=0.17 deg C per 1 mW, enabling 600 mW to be dissipated.

In some circumstances, the improved dissipation is offset by the fall of h_{FE} at higher collector currents. A higher drive power then becomes necessary. It may well be more economical to operate output transistors in parallel to achieve a given output power than to use a heat sink and more powerful drive circuits.

STABLE VOLTAGE REFERENCE SOURCE

A ring-of-2 reference circuit that provides two identical or dissimilar reference voltages is described here. The circuit is shown in Fig. 13-10.

The ZTX303 transistors are connected as diodes, with the reverse-biased emitter junction acting as a zener diode. (In Fig. 13-10, only one ZTX303 is shown.) This reference source stabilizes the base-emitter potential of a transistor, causing a constant collector current. The current flowing across a

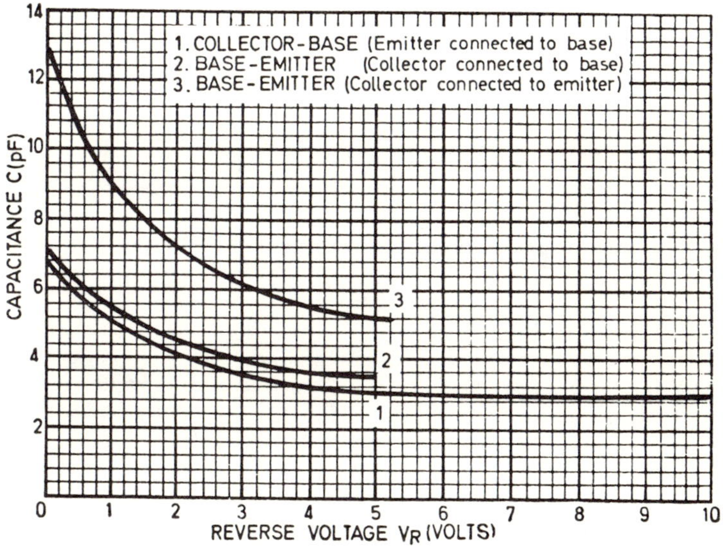

Fig. 13-9. Graph of the variation of capacitance with reverse voltage.

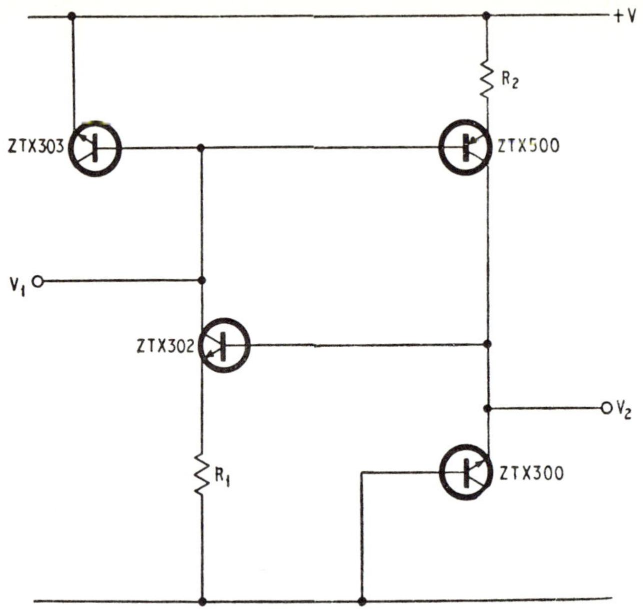

Supply Voltage	V_{out1} (via a 3.2 : 1 probe)	Stabilization Ratio S
11.0	1.6507	—
12.0	1.7888	—
13.0	1.9065	—
14.0	2.1066	—
15.0	2.1088	100,000 : 1
20.0	2.1088	
25.0	2.1088	

Fig. 13-10. Ring-of-2 reference circuit.

second reverse-biased junction stabilizes the base-emitter voltage of the second transistor, whose collector current is the constant-current source for the first simulated zener diode, completing the ring. Two reference sources are thus available. The permitted range of supply voltage varies from the base-collector breakdown voltage of the transistor to a minimum voltage given by V supply min equal to or greater than

$$V_{Z2} + V_{C1(sat)} + (V_{Z1} - V_{be1})$$

where VZ1 and VZ2 are the base-emitter breakdown voltages, $V_{C1(sat)}$ is the saturation voltage of the first transistor and $V_{Z1} - V_{be1}$ is the voltage across resistor R1. With the circuit shown, V_{Z1} approximates V_{Z2} 8V, V_{C1} is typically 0.2V, and V_{be} is typically 0.8V. The transistor will saturate with a supply voltage of 8 + 0.2 + (8—0.8) = 8.2 + 7.2 = 15.4V. Therefore, the minimum supply voltage must be above this value.

The stabilization ratio ($S \Delta V_{in} \Delta V_{out}$) was measured for a range of supply voltage, with R1 and R2 selected to give 1 mA collector current in each transistor (R1=R2=6.8K). Dissimilar collector currents had an adverse effect on stability.

Because the supply current is the sum of the collector currents, the circuit may be considered as a 2-terminal device that is across and using the supply terminals; in this configuration, it can be used as a constant-current source. For a typical line current of 2.5 mA, a change of applied voltage of 5V will produce a current change of 3 uA or 0.1 percent.

DC AMPLIFIER

In Fig. 13-11, Tr1 and Tr2 are a matched pair of ZTX300 transistors ($h_{FE1} = h_{FE2}$ plus or minus 10 percent at $I_c = 100$ uA). Both transistors are held together in a "DELTA" model 203 (TO-18) sink, or are mounted in a Redpoint heat sink, in order to maintain them both at a common ambient temperature. Transistors Tr3 and Tr4 are matched ZTX302 types, which are held together in a similar manner. Potentiometer VR1 sets the output to zero voltage for zero input.

Typical performance characteristics of the amplifier are as follows:

Maximum output voltage: plus or minus 11V at plus or minus 10 mA

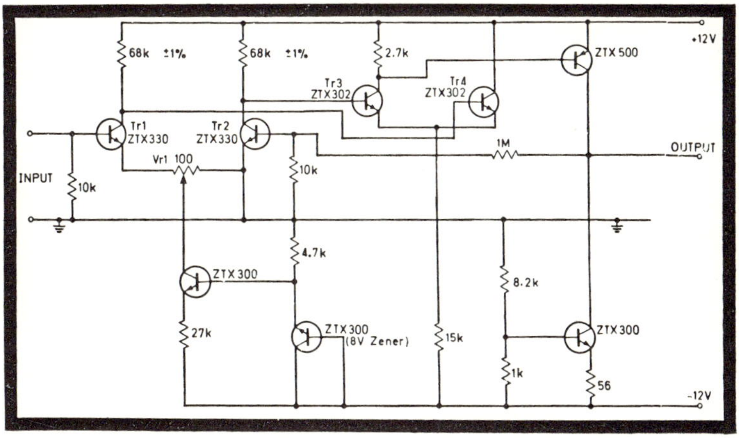

Fig. 13-11. The dc amplifier.

Minimum load for full voltage swing: 1.1K
Nominal voltage gain: 100
Bandwidth (3 dB): 800 kHz
Noise referred to input: less than 200 uV peak to peak
Input impedance: 10K
Temperature drift: Approximately 25 uV referred per deg C to input)

The circuit must be shielded from drafts to prevent a temperature differential arising between Tr1 and Tr2 that could give rise to a large drift of output voltage.

CHOPPER DC AMPLIFIER USING ZTX300

An alternative method of dc amplification is to chop the applied dc signal to produce a square wave of amplitude corresponding to the dc level. This can then be amplified in an ac amplifier that may have a large gain, but no dc drift problems. The output of the amplifier is fed to a demodulator, which restores the signal to its dc component. The chopper and demodulator are usually synchronized.

The ZTX300 transistor has a very low offset voltage when operating in the inverse mode; that is, with the collector and emitter reversed. The typical value of the offset voltage developed across a 1.5K load for a base input current of 500 uA is about 0.25 mV. Typical slope resistance for a base input current of 500 uA is 10 ohms for both positive and negative inputs.

The Shunt Chopper

The voltage to be amplified is modulated by the chopper. The resultant square wave is amplified, then synchronously demodulated, to give a dc output with the same polarity as that at the input (see Fig. 13-12).

Single-Transistor Chopper

See Fig. 13-13. For good linearity, V_{in} should be much greater than the offset voltage of a transistor. That is $V_{in} = 10$ mV for reasonable linearity.

BALANCED CHOPPER AMPLIFIER-MICROVOLT—MICROAMP METER

The chopper transistor offset voltage can be balanced out with that of another, similar transistor, and voltages below 100

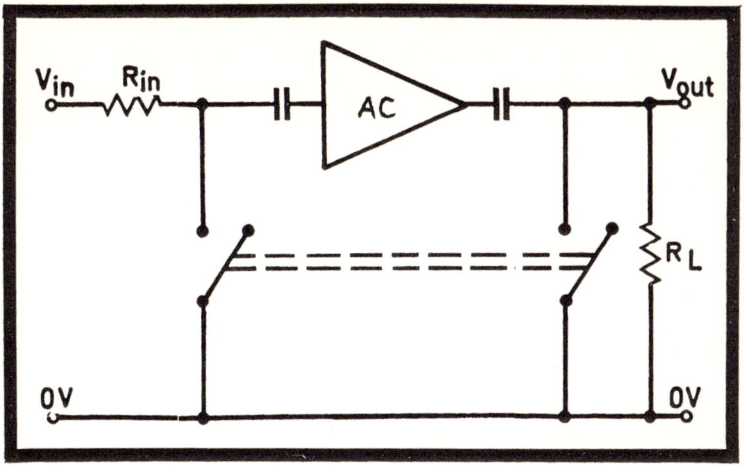

Fig. 13-12. Shunt chopper block diagram.

uV can be measured accurately. In addition, when the transistors are used in a balanced circuit, temperature drift is reduced as both transistors drift together. See Figs. 13-14 and 13-15.

The meter used should be a 100 uA-0-100 uA, center-zero, moving-coil type, with an internal resistance of less than 1K.

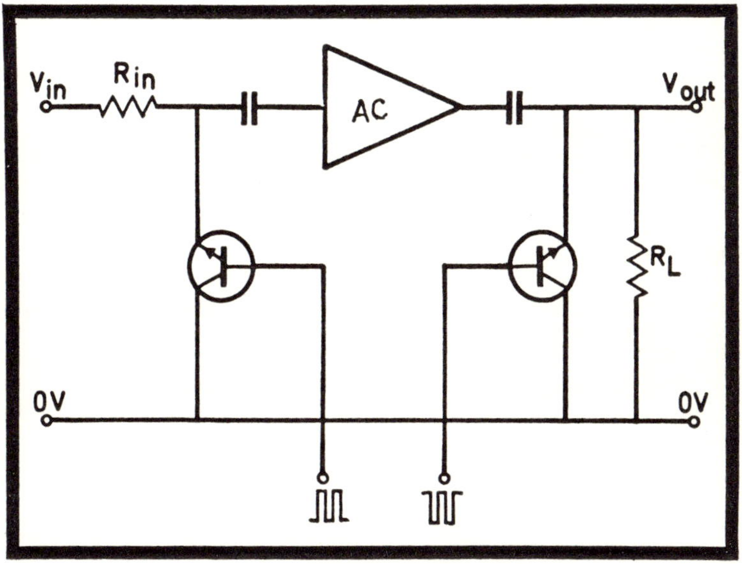

Fig. 13-13. Single-transistor chopper.

Setup

Adjust VR1 to give zero reading on the meter when the switch is in the 1V position and with zero input.

With the range switch at 100 uV and VR2 at approximately midposition, feed in 100 uV from an impedance source of less than 10 ohms, and adjust VR3 for full-scale deflection.

Adjust VR2 until full-scale deflection is also obtained for an input with the range switch in the 1V position.

For accuracy in the 100 uV-100 nA range, the source resistance must be less than 100 ohms.

Because of the high sensitivity of the ac amplifier, the whole unit (excluding power supply) should be enclosed in a metal box grounded to the circuit at the input only. The metal box will also shield Tr1 and Tr2 from drafts and improve the short-term temperature stability of the amplifier. All wiring must be as short as possible.

A low ripple power supply must be used. A suitable circuit is shown in Fig. 13-16. The output is set to 12V by adjusting VR1. Ripple is approximately 15 mV p-p.

Specification

Supply voltage is 12V at approximately 35 mA. (Initial current upon switching on is considerably higher than this for a few seconds because of the long time constants involved.)
AC amplifier:
Power gain: 86 dB typical
Bandwidth: 20 Hz to 150 kHz (minus dB)
Noise less than 10 uV
Chopping frequency: 1.5 kHz (approximate)
Input impedance of voltage ranges: 10M per volt
Voltage drop on current ranges: 200 uV
Linearity, excluding meter: better than 3 percent

The final accuracy of the instrument will depend upon the accuracy with which the initial adjustments were made.

TEMPERATURE DRIFT

The chopper type of amplifier has a vastly superior temperature-drift figure in comparison with the direct-coupled differential amplifier.

For the best temperature-drift performance, Tr1 and Tr2 should be mounted together in a Redpoint type 92-DC heat sink to keep them both at the same temperature.

No attempt was made to match Tr1 and Tr2, as the values of offset voltage and slope resistance were fairly constant for the several specimens tested.

The temperature drift was found to be approximately 1 uV per deg C, referred to the input.

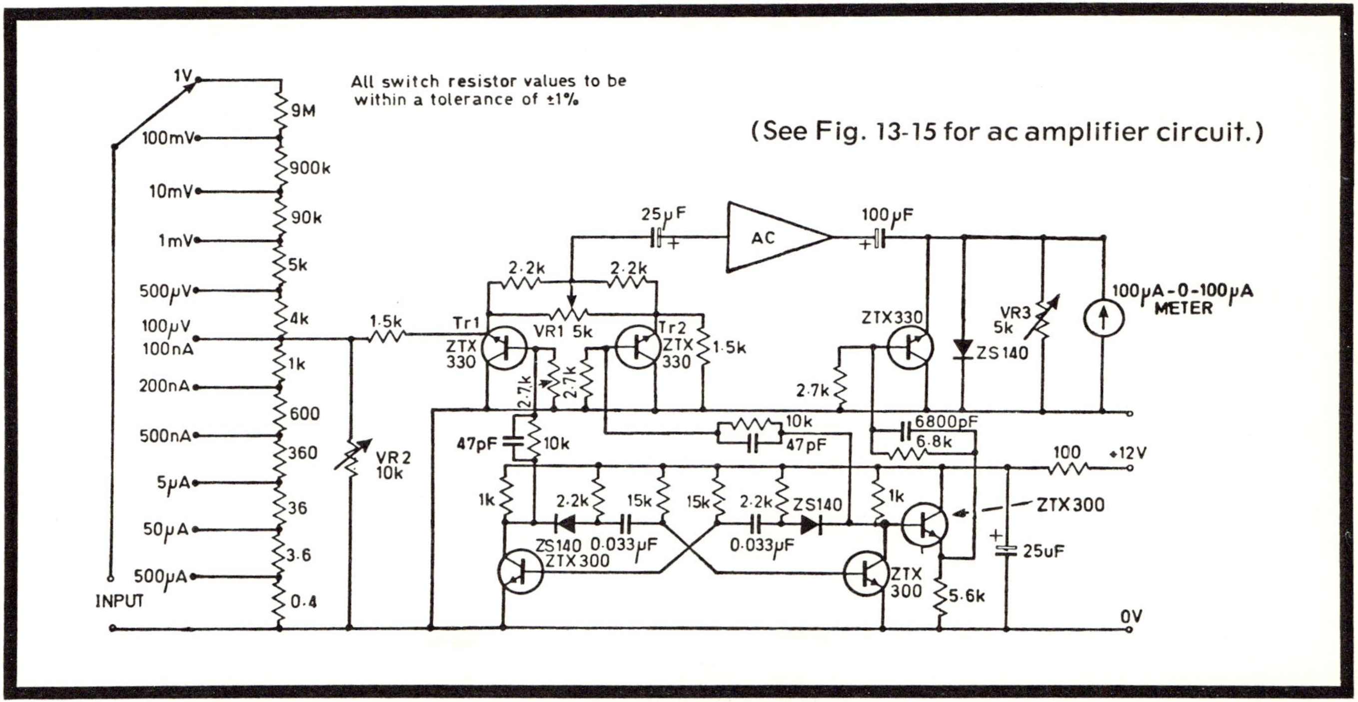

Fig. 13-14. Balanced chopper amplifier.

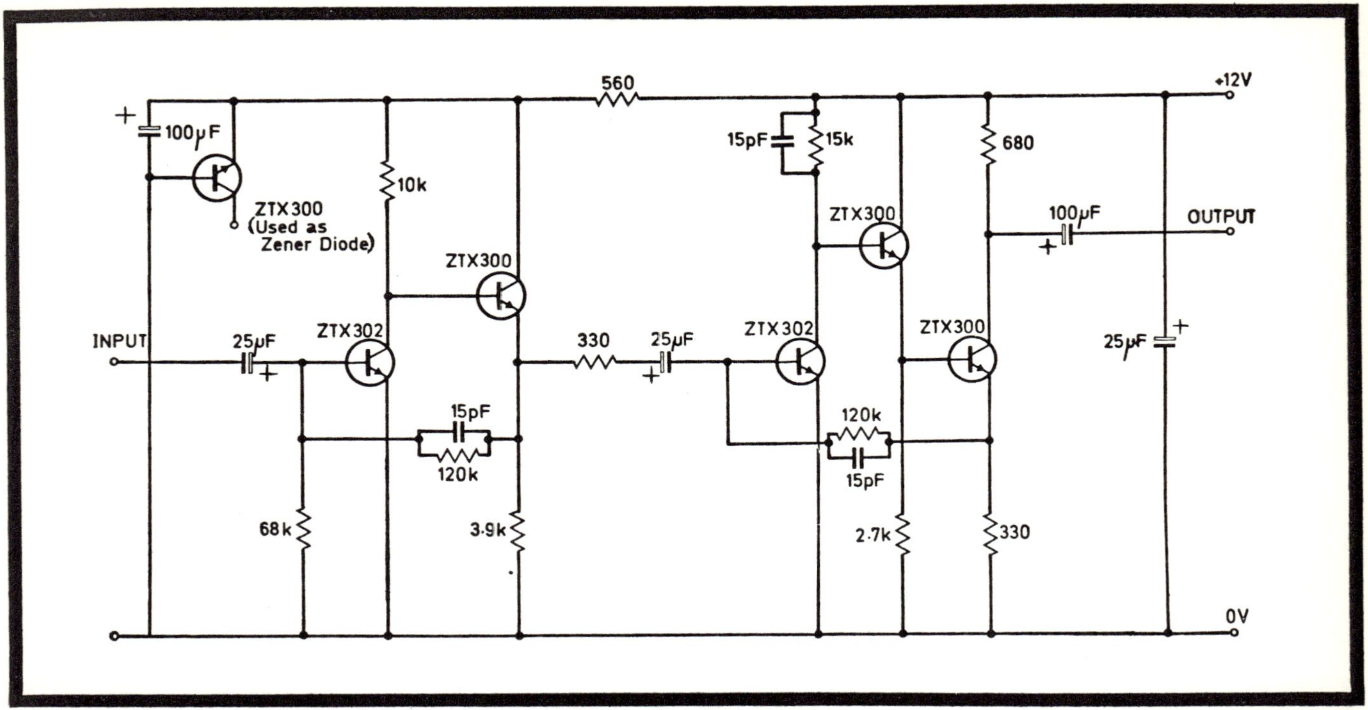

Fig. 13-15. The ac amplifier for Fig. 13-14.

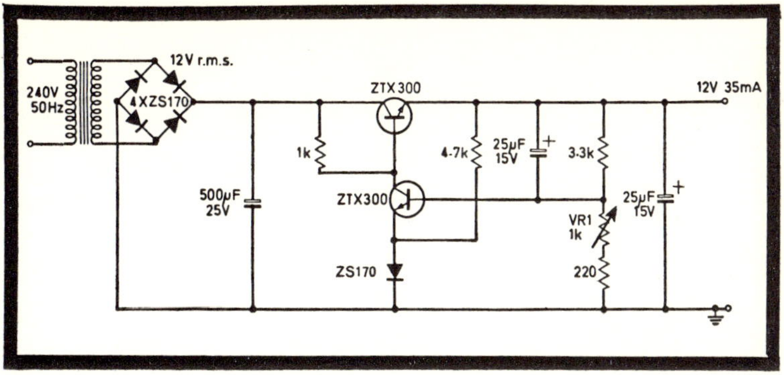

Fig. 13-16. Power supply for balanced chopper amplifier.

Part II

Complementary Symmetry Audio Power Amplifiers

Chapter 14

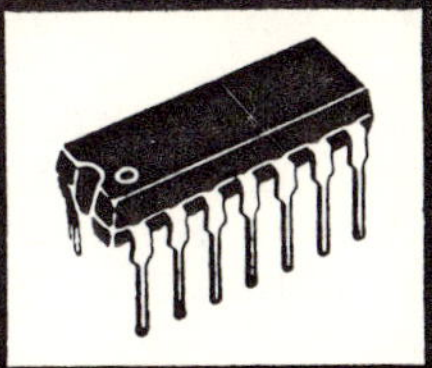

Complementary Symmetry Power Amplifier Design

Most output circuits of audio amplifiers used today are formed by a series connection of two transistors of the same polarity (quasi-complementary) or of opposite polarity (complementary) across the power supply. The load is coupled directly to the midpoint (split operation) or coupled through a capacitor (single-supply operation).

Complementary symmetry gives by far the best circuits for stability and simplicity. Therefore, complementary pairs are widely used for low output power with both germanium and silicon transistors. Power has been limited to a few watts by the lack of silicon pnp and germanium npn high-power transistors; however, the design of complementary audio amplifiers up to 100W is now possible through the introduction of a broad range of Motorola silicon pnp power devices.

The true complementary configuration is represented in Fig. 14-1. Somewhat different connections are usable, but in any case the impedance between emitter and base for both

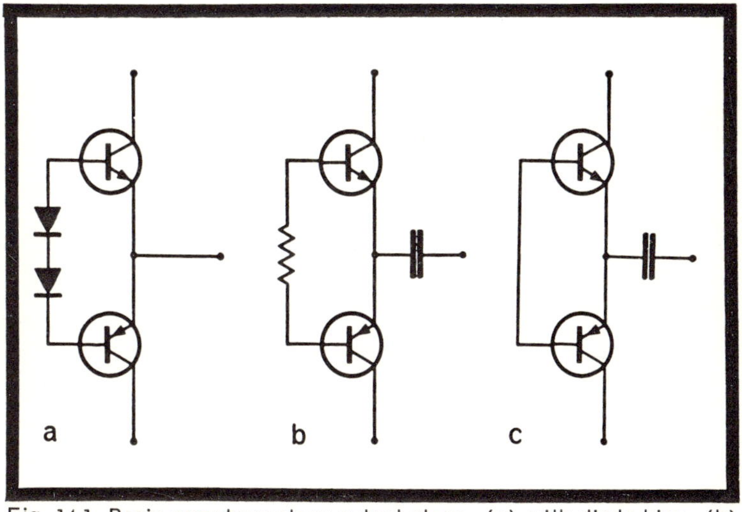

Fig. 14-1. Basic complementary output stage: (a) with diode bias; (b) with resistor bias; (c) unbiased.

transistors is low, indicating poor stability. If diodes are used for biasing, a fairly constant quiescent current vs supply voltage and temperature can be achieved. The parameter responsible for the static stability is the leakage current I_{CBO} when perfect matching of all other characteristics is assumed.

The configurations of Figs. 14-16b and 14-16c are used for inexpensive circuits. Crossover distortion in the latter circuit must be eliminated by negative feedback; a higher total open loop is thus required. A disadvantage of the complementary configuration might be the requirement for a class A driver. The dissipation of the driver stage could be quite high in high-

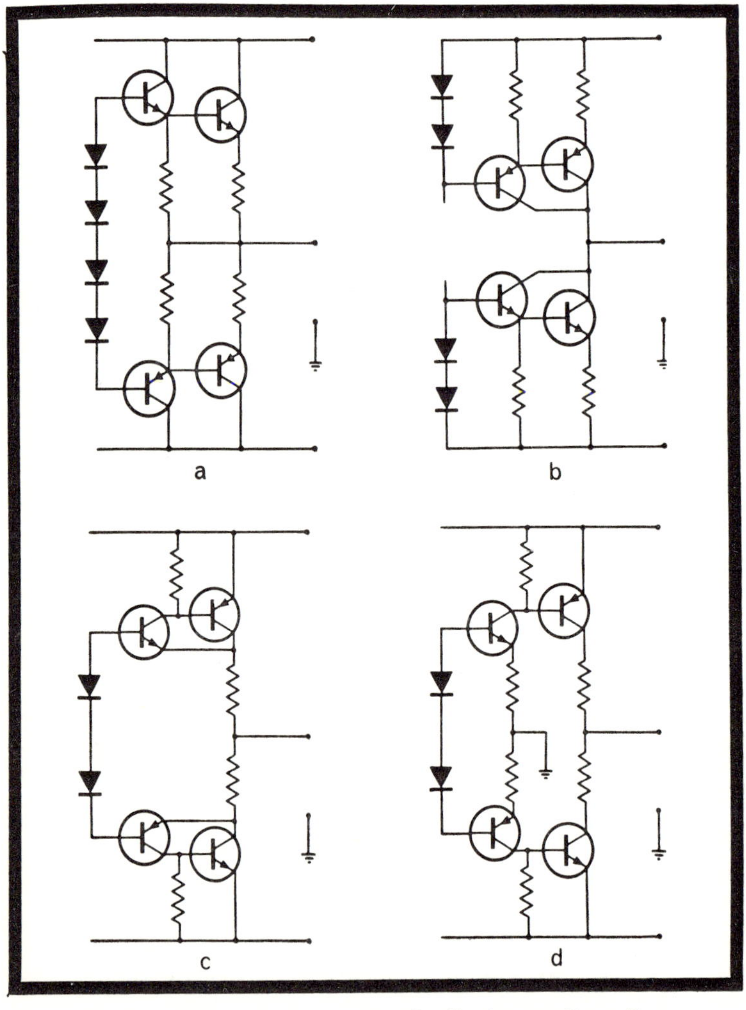

Fig. 14-2. Basic complementary Darlington configurations.

power amplifiers. The remedy is to use some form of Darlington configuration, as shown in Fig. 14-2.

The circuit of Fig. 14-2a is a straightforward solution to the problem, and is practical. It does, however, present a few biasing problems. The base-to-emitter **on** voltage at the threshold of conduction is significantly different between the output and driver devices. The required bias voltages (the sum of the driver and output V_{be}) do not match well with the intrinsic forward characteristic of the bias diodes. Some form of adjustment is necessary—this tends to unbalance the bias tracking and compensation sought by the use of diodes.

Figure 14-2b offers another interesting possibility, but the practical problems are similar to those of Fig. 14-2a. Furthermore, the bias diodes in this configuration shunt the desired drive signal to ground and seriously load the source. Figure 14-2c offers a connection in a cascaded complementary set that functions as an emitter follower. This basic arrangement has distinct advantages in that the diodes and driver transistors commonly match well in forward characteristics, and the output devices are completely slave driven. The stability and linearity of the emitter follower are thus available, and the bias is fully compensated by matching diodes.

Figure 14-2d offers an additional combination that has the advantage of full diode bias with the power gain of two cascaded common-emitter stages. Thermal stability is, however, more precarious than that of the other circuits because no direct bias-current feedback is provided.

The circuit of Fig. 14-2c is the most interesting, then, as an inherently linear, stable configuration that can offer the ease of a compensated and adjustment-free bias scheme. An illustration of the order of magnitude of improvement experienced with the diode bias circuit can be seen in Fig. 14-3. The curves indicate the drift of idling current in the power transistors as the nominal power-supply voltage is varied plus or minus 10 percent on a 50W experimental amplifier. The comparisons are between a quasi-complementary circuit (Fig. 14-4) with diode or series-resistor bias adjustment and the full complementary circuit of Fig. 14-2c with dual diode bias. The latter is the least sensitive to such supply-voltage variations.

The situation becomes similarly more favorable for the full complementary scheme when ambient operating temperature variations are taken into account. Also, the form of the curves is about the same for this situation (see Fig. 14-5).

The diodes used in the amplifier of Fig. 14-2c are matched in their forward characteristic to the transconductance (V_{be} vs I_c) of the driver devices in the idling condition. The spread is basically within plus or minus 3 percent. The resulting tolerance of idling current for the power devices is small enough so that no adjustment of bias is necessary.

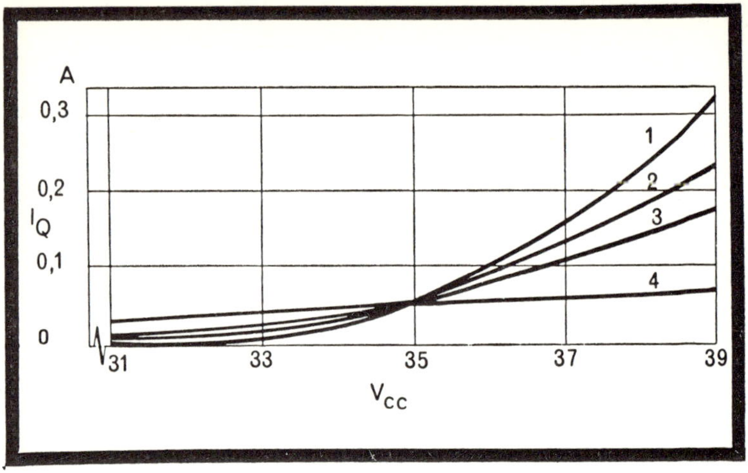

Fig. 14-3. Sensitivity of idling current bias in output power transistors to variations in power supply voltage for quasi complementary-symmetry connection with common bias circuits, and full conjugate complementary connection in experimental 50W amplifier. Bias set to 50 mA at V_{CC} + 35V at start.

1. Quasi-complementary: resistive bias set only
2. Quasi: resistor + 1 diode
3. Quasi: resistor + 2 diodes
4. Conjugate complementary: dual diode bias only

OUTPUT STAGE

Two different types of output configuration will be considered: one using complementary single transistors for low-power output, the other using complementary compounds for higher power (Figs. 14-6a and 14-6b). The choice between the two configurations is primarily determined by the dissipation of the driver, which is inversely proportional to the total current gain of the output devices and directly proportional to the output power. The dissipation of the driver must be such that the total amplifier efficiency is not significantly degraded.

Distortion

Fig. 14-7 shows the transfer curves (collector current, I_C, vs input voltage, V_{in}), for the common-emitter, common-base, and common-collector configurations. The common-emitter circuit always has poor linearity; the common-base circuit is only linear with current drive; the common-collector (emitter-follower) circuit exhibits nonlinearity with current drive. With voltage drive, however, it is quite linear. Input and output voltages differ then by V_{be}, which changes very little with collector current.

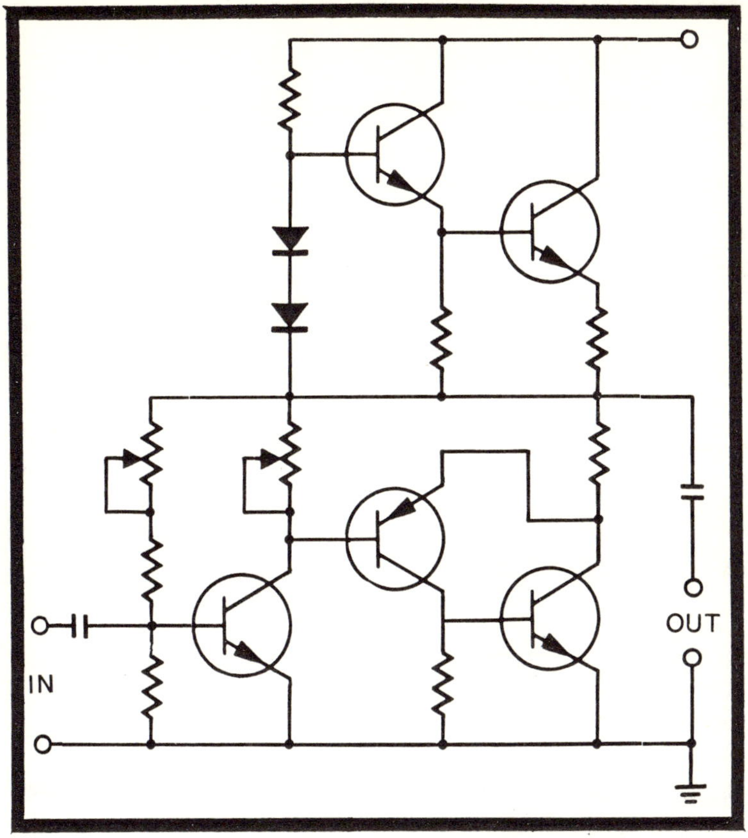

Fig. 14-4. Basic quasi complementary symmetry.

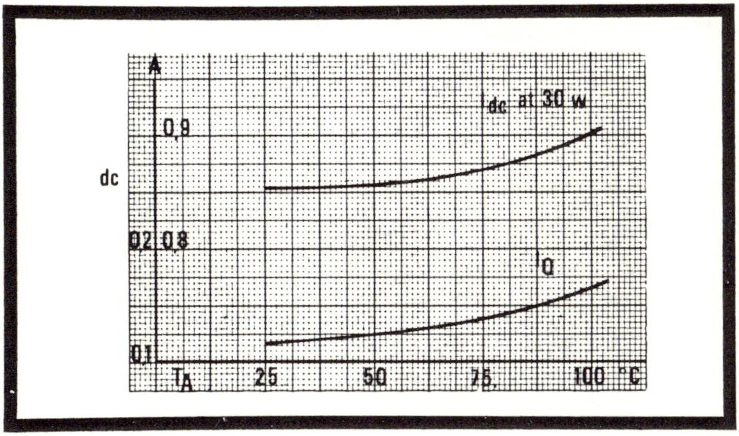

Fig. 14-5. Quiescent current as a function of temperature.

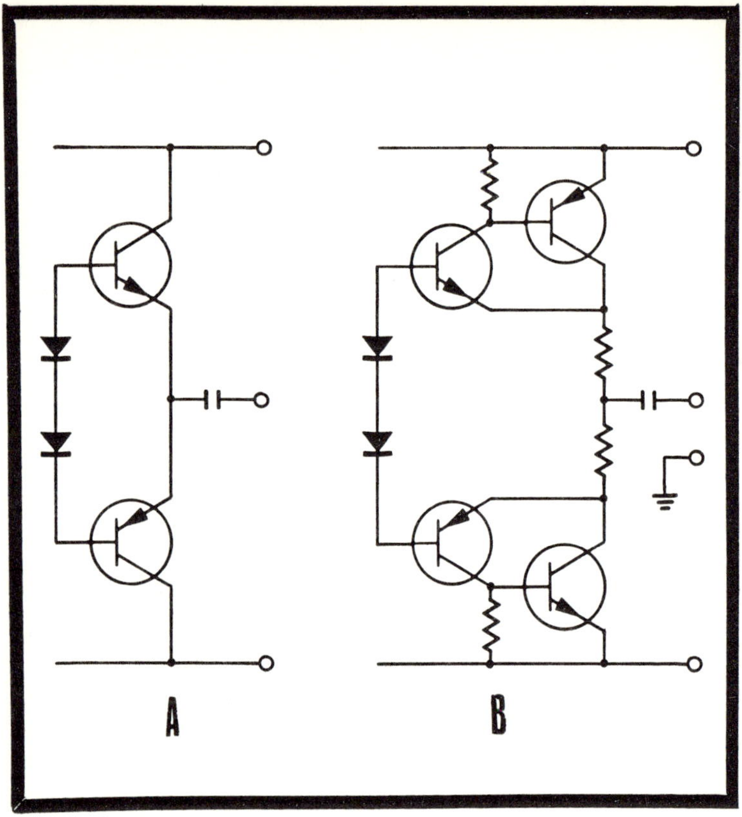

Fig. 14-6. Basic output-stage circuit configurations: (a) Single; (b) Compound.

Mismatch

Mismatch between the units of a pair is a cause of distortion. This can be greatly minimized in the emitter-follower configuration, as mentioned above. Furthermore, the negative feedback—as high as 45 dB for the 60W amplifier—greatly reduces total harmonic distortion.

The device suggested for each schematic given later will provide excellent low-distortion operation for all of the parameter values down to or up to the specified limits of current gain without any matching of the pairs.

Maximum Output Swing

The maximum swing is slightly smaller than the total supply voltage. With the simple complementary pair of Fig. 14-8, the situation is as described below.

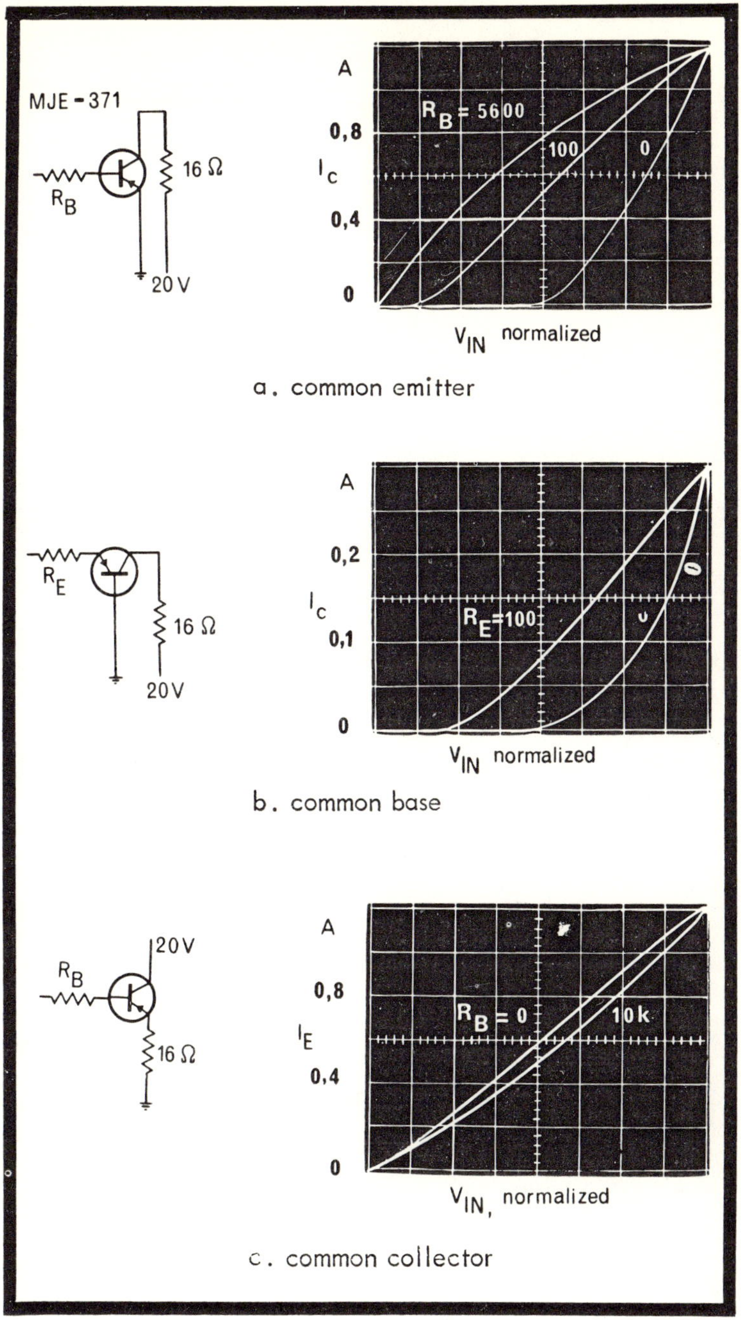

Fig. 14-7. Linearity of the three basic transistor configurations.

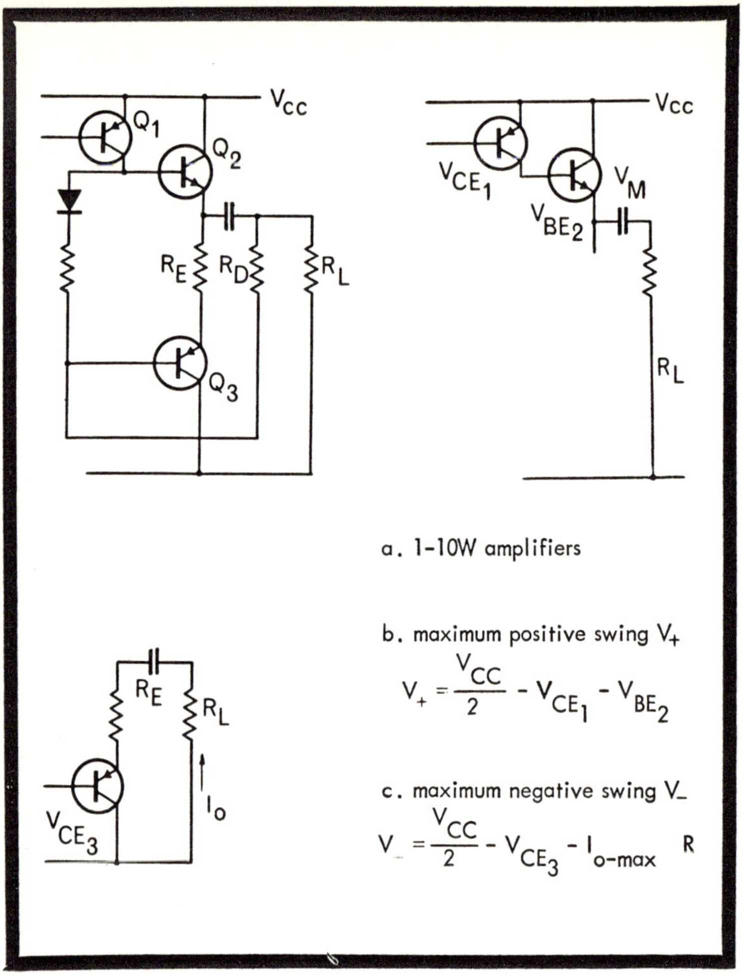

Fig. 14-8. Maximum output swings.

For the 10W amplifier (described below), the total output swing will be 41 — (0.5 + 0.7) — (1 + 1.12) = 37.68V, corresponding to a power of 11W typically. The midpoint voltage will have the value:

$$VM = \frac{41 - (0.5 + 0.7) + 1 + 1.12 \quad 20.8V}{2}$$

As can be seen from Fig. 14-8, sufficient drive will be supplied to npn Q2 because its base is connected to V_{CC} through driver Q1. But the base current for the pnp is limited by R_D. The value of R_D will be then determined as a function of the minimum current gain of Q3. Furthermore, to increase the base drive of Q3, bootstrapping (i.e., returning R_D to the hot terminal of the load) is used.

When the driver is an npn and is connected between Q3 and ground, R_D is determined by the minimum current gain of Q2. For the compound configuration, the maximum swing is similarly limited.

Output Device as a Function of Output Power Load

Table 14-1 is provided to facilitate selection of the output devices as a function of required output power and load. Normally, the indicated peak current I_{pk} should be greater than the current at which the minimum current gain is guaranteed. A gain that is too low, especially for Q3 in Fig. 14-8, could lead to an excessive dissipation in the class A driver.

Safe Operation

Safe operation will be examined through analysis of the load lines. This will be done for the 10W amplifier using the test circuit of Fig. 14-9.

The first method consists of the inspection of the Lissajous figure obtained as indicated in the figures discussed below.

Figure 14-10 shows the load line for a resistive load, and Fig. 14-11 shows the load line for a capacitive load; an inductive load would give similar results. In the case of audio amplifiers connected to a loudspeaker, the load becomes inductive over about 1 kHz. However, because of the high impedances involved, this case does not have to be considered for safe operation.

The capacitive load is formed essentially by the coupling capacitor at frequencies below about 100 Hz, depending upon the speaker and capacitor.

A device will work safely as long as its load line falls within the corresponding Safe Operating Area (SOA). The SOA to be considered is the one specified at a time equal to one-half cycle of the test frequency. For instance, at 1 kHz, the SOA for 1.5 msec or very low frequencies, such as 50 Hz or less, the corresponding SOA will be the dc area.

With a loudspeaker as a load, the worst case occurs at low frequencies caused by the decrease of loudspeaker impedance by the coupling capactor (in single-supply configurations). To simulate such conditions, Figs. 14-11 and 14-12 were obtained at 60 Hz with a load of 8 ohms in series with 200 uF.

The shape of the figures depends upon the input voltage for reactive loads only. Figures 14-10 and 14-11 were obtained

TABLE 14-1.

f_o	f_1	f_3	f_4	ϕ
0	1	∞	∞	0
0.001	0.99	15.7	6.96	8°
0.003	0.98	10.5	3.5	11°30'
0.008	0.96	7.85	1.98	16°15'
0.015	0.94	6.3	1.68	20°
0.022	0.92	5.7	1.55	23°
0.031	0.90	5.1	1.46	26°
0.057	0.85	4.15	1.29	31°45'
0.075	0.82	3.77	1.24	35°
0.088	0.80	3.57	1.21	37°
0.124	0.75	3.2	1.12	41°30'
0.158	0.70	2.98	1.075	45°30'
0.200	0.65	2.75	1.036	49°30'
0.245	0.60	2.55	1.01	53°
0.293	0.55	2.4	0.979	56°30'
0.344	0.50	2.3	0.954	60°
0.393	0.45	2.2	0.93	63°15'
0.453	0.40	2.1	0.907	66°30'
0.513	0.35	2.0	0.884	69°30'
0.574	0.30	1.9	0.866	72°30'
1	0	1.57	0.785	90°

ϕ = half angle of conduction

$f_o = \sin\phi - \phi\cos\phi$

$f_1 = \cos\phi$

$$f_3 = \frac{\pi(1 - \cos\phi)}{2f_o}$$

$$f_4 = \frac{\sqrt{\pi}\sqrt{2\phi - 3/2 \,.\, \sin 2\phi + \phi\cos 2\phi}}{2\sqrt{2}(\sin\phi - \phi\cos\phi)}$$

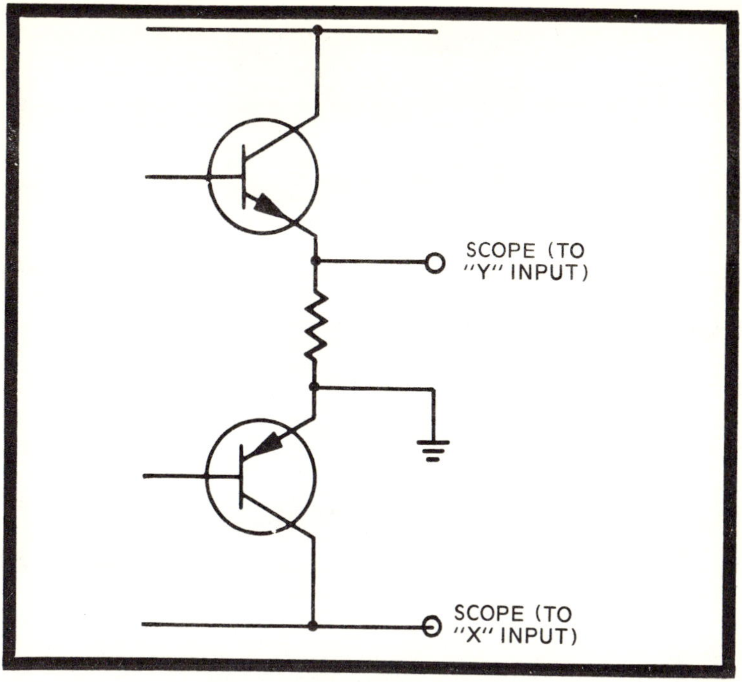

Fig. 14-9. Load-line test circuit.

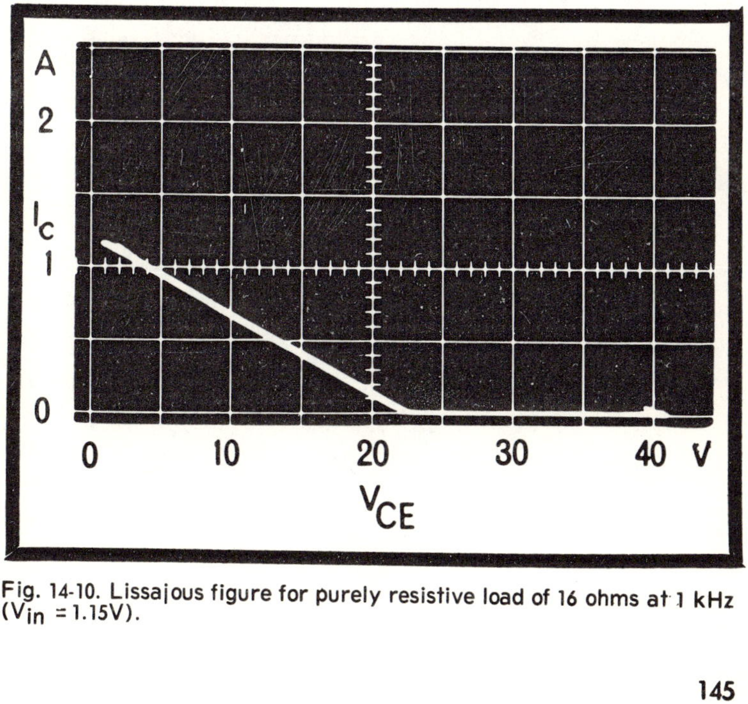

Fig. 14-10. Lissajous figure for purely resistive load of 16 ohms at 1 kHz (V_{in} = 1.15V).

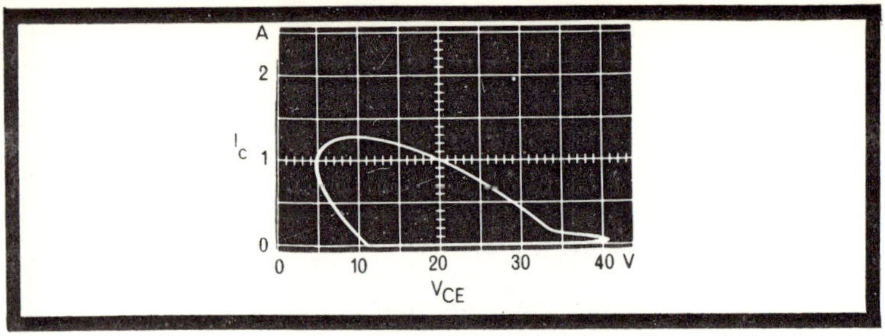

Fig. 14-11. Lissajous figure for capacitive load (60 Hz) with V_{in} 1.15V.

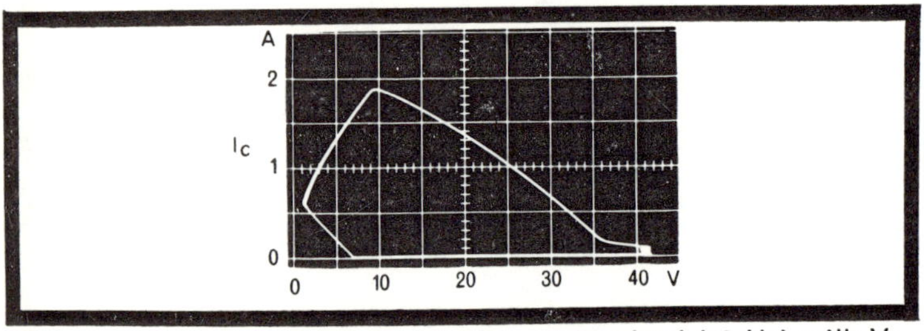

Fig. 14-12. Lissajous figure for purely resistive load (60 Hz) with V_{in} = 2.3V.

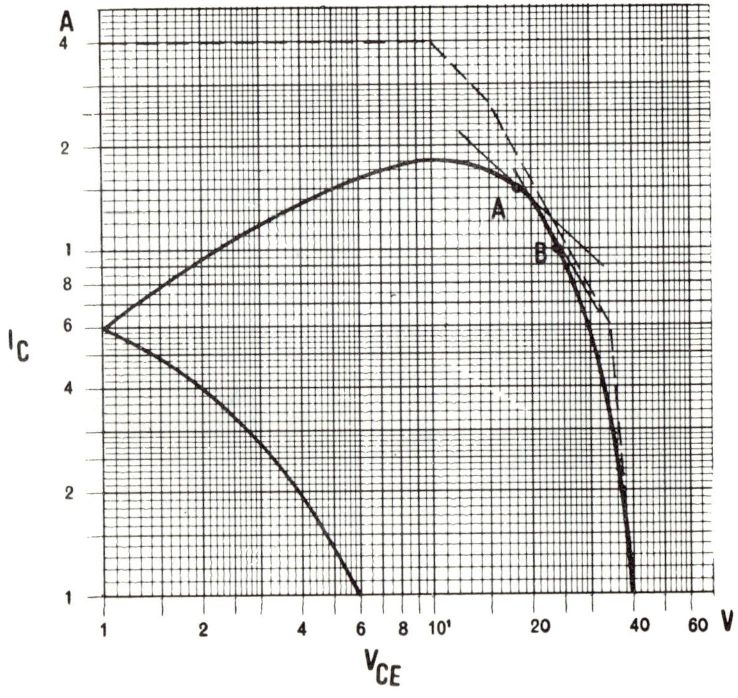

Fig. 14-13. Load line and permissible area (refer to Figs. 14-12 and 14-16).

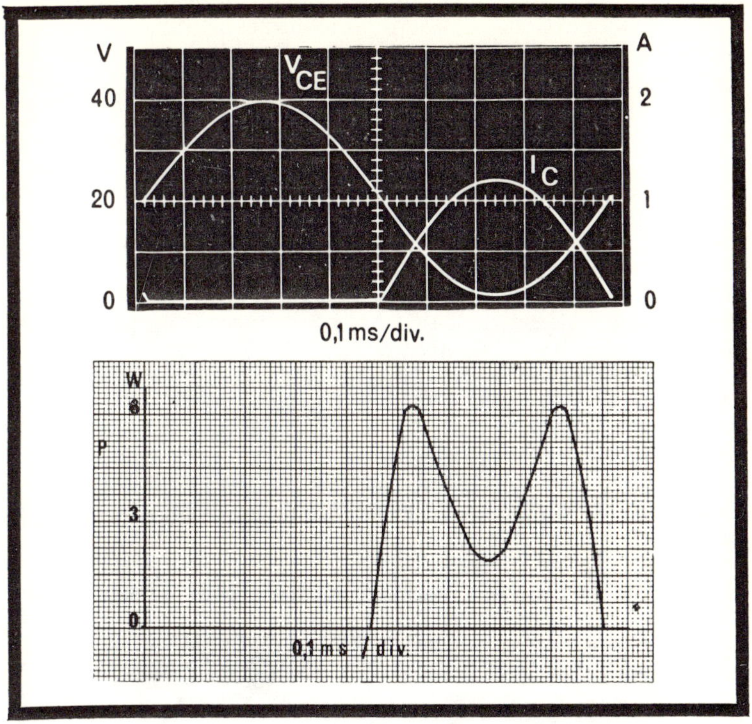

Fig. 14-14. Power vs time for the 10W amplifier with resistive load (V_{in}= 1.15V).

with an input voltage of 1.15V just before clipping. Figure 14-12 was obtained with 2.3V in conditions of saturation.

When plotted in the SOA of an MJE371 (or 2N5193) transistor, as shown in Fig. 14-13, the load line does not cross the dc area, and operation is safe. This method is simple and easy to use. However, the area enclosed by the load line is much smaller than the permissible one, indicating that this test is too conservative.

Because the curves of an SOA are obtained with rectangular pulses of different durations and energy contents, a more accurate analysis will make use of the actual power pulses (as described below).

Figures 14-14 and 14-15 refer to the previous case of resistive load (Fig. 14-10). The power curve is obtained by multiplication of V_{ce} by I_c. Two power peaks of equal amplitude appear in one cycle. Their energy content decreases as the input voltage increases; therefore, total dissipation lowers as the transistors enter saturation. Figure 14-16 refers to the previous case of capacitive load. The correlation that must first be established between any actual power-pulse form and an equivalent rectangular pulse is the equality of their energy contents. Then, to determine the **pulse amplitude**, two different considerations must be made corresponding to the two types of power limitation discussed above. For power-dissipation limitation, the power-pulse amplitude is the peak power as determined by Fig. 14-16. It is also point A in Fig. 14-13, where the slope of the tangent to the load line is equal to (—1) in double-logarithmic scale. For second-breakdown limitation, the highest amplitude is given by point B in Fig. 14-13, where the slope of the tangent is parallel to the next specified SOA limit.

At point A, the **equivalent rectangular pulse** has the following characteristics: 20V x 1.35A x 2.85 msec. At point B, the characteristics are: 25V x 1A x 3.1 msec. Both points are plotted in Fig. 14-17, as shown.

The device is second-breakdown-limited if the case temperature is equal to 25 deg C. To determine the case temperature at which the limitation would be imposed by the dissipated power, the calculation can be carried out as indicated in Fig. 14-18. At a case temperature of 110.5 deg C, the junction temperature would reach 150 deg C at the end of each pulse.

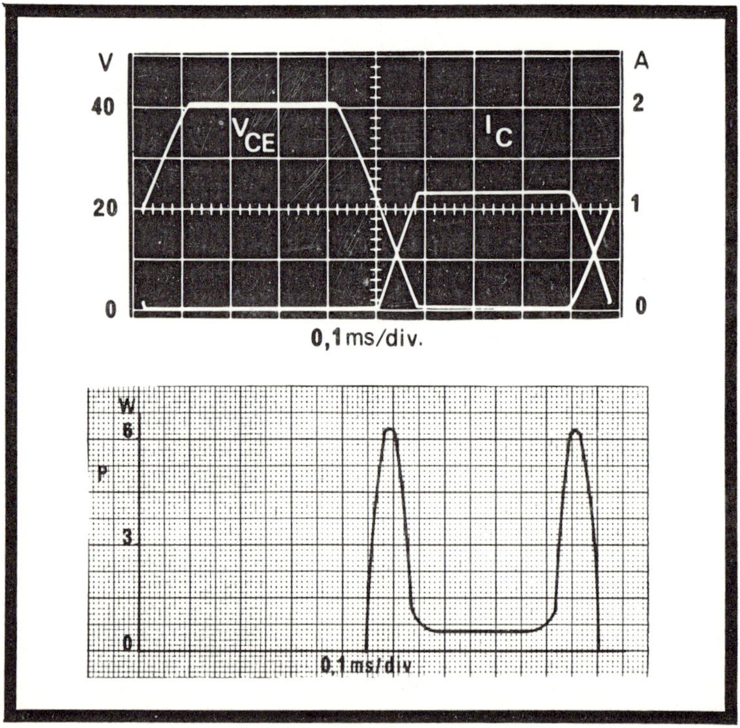

Fig. 14-15. Power vs time for the 10W amplifier with resistive load (V_{in}= 2.3V).

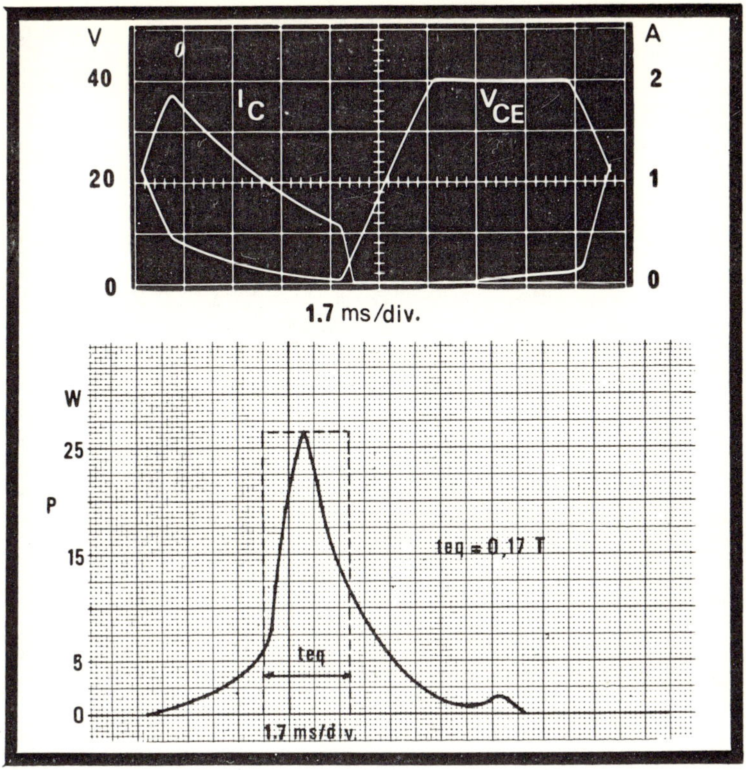

Fig. 14-16. Power vs time for the 10W amplifier with capacitive load (V_{in}= 2.3V).

The 10W amplifier has been examined. The safe operation of other amplifiers with single supply can be investigated in the same manner. Amplifiers with split supply are similar in principle; however, these will be examined later in conjunction with the protection circuit.

DRIVER STAGE

For noncompound, low-power amplifiers, the driver (Fig. 14-19) includes two transistors, Q2 and Q3, in cascade, working as individual common-emitter stages. The load resistance R_D and the dissipation of Q3 in the quiescent condition are determined by the maximum base current passing through Q5 during peak signal excursions:

$$RD = \frac{V_{cc}}{2I_{pk\ out}} \times h_{FEmin} \text{ or } R_L \times h_{FEmin}$$

where I_{pk} out is the peak current at the output, h_{FEmin} is the minimum current gain of Q5, and Rl is the load resistance.

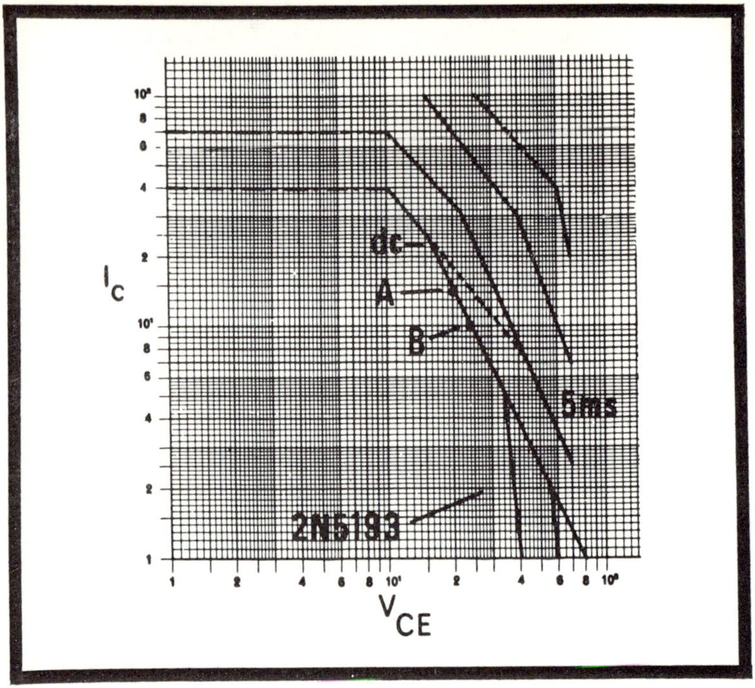

Fig. 14-17. Safe operation area on the MJE371 (2N5193).

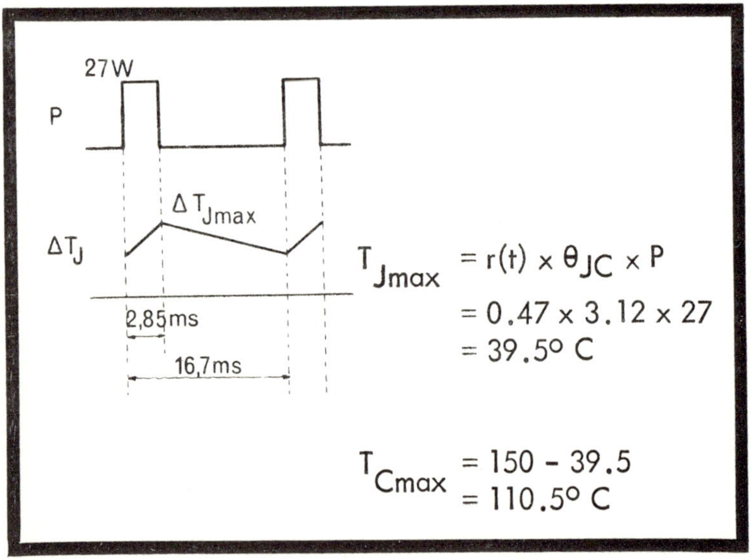

Fig. 14-18. Maximum case temperature for 10W amplifier.

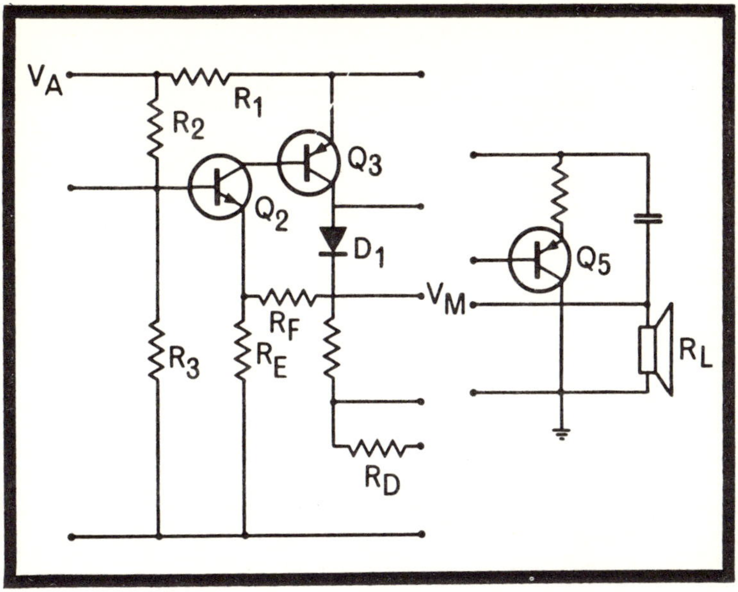

Fig. 14-19. Driver stage for low-power amplifiers.

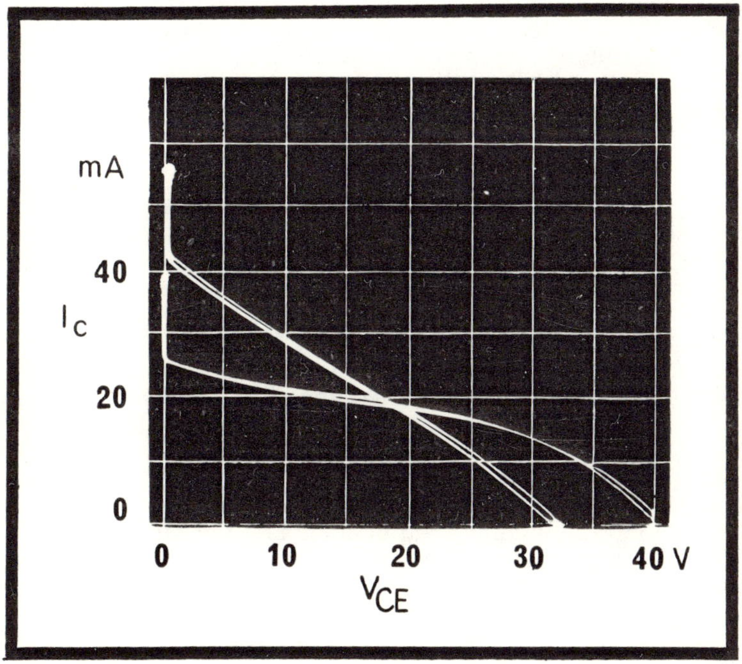

Fig. 14-20. Load line of Q3.

The power dissipated in Q3 for the quiescent condition is approximately equal to:

$$P_{D3} = \frac{V_{cc}^2}{4 R_L h_{FEmin}} \quad \text{or} \quad \frac{2P_{out\,max}}{h_{FEmin}}$$

For the 10W amplifier with a load of 16 ohms, I_{pk} is equal to 1.12A. The MJE371 is specified at h_{FEmin}=40 at Ic=1A; thus, based upon the normalized gain curve, h_{FEmin} at 1.12A will be 38. Then, with V_{cc} equal to 41V,

$$R_D = \frac{41}{2 \times 1.12} \times 38 = 700 \text{ ohms}$$

and

$$P_{D3} = \frac{2 \times 10}{40} = 0.5\text{W}$$

The dissipation of Q3, P_{D3}, will increase when the output power increases, and will require higher current gains, which is precisely a feature of the compound configuration. The total closed-loop dc and ac voltage gain of the amplifier (excluding the preamplifier) is determined by the ratio of feedback resistors R_F and R_E, or

$$(R_F + R_E)/R_E.$$

The **open loop-gain** is determined only by Q3, as Q2, Q4, and Q5 have no voltage gain.

The load line of Q3, calculated earlier, was shown to be dependent upon the current gain of the output transistors. For very high gains, it is nearly horizontal. The current passing through Q3 is, then, nearly constant. Figure 14-20 shows the measured load line of Q3. The load line obtained without bootstrapping is shown in the same figure for comparison. Note that the maximum voltage excursion is reduced.

The second transistor (Q2) used in the driver (Fig. 14-19) has a voltage gain near unity. It can be thought of as a voltage comparator between the supply voltage V_{cc} and the center voltage V_M, maintaining V_M at about half the supply voltage. To fulfill this condition, the voltage dividers must be chosen so that

$$\frac{R3}{R1 + R2 + R3} = \frac{R_E}{2(R_E + R_F)}$$

It is advisable that all of these resistors have a 5 percent tolerance so that V_M will also change as a function of the current gain of Q3. In practice, it will be further influenced by changes in the beta of Q2, but to a lesser extent.

The above equation must be corrected, because of the additional voltage drop at R1 caused by the preamplifier current. More accurately,

$$V_M = \frac{R_E + R_F}{R_E}\left[V_{cc}\,\frac{R3}{R1 + R2 + R3} - V_{be\,Q2}\right]$$

For the 10W amplifier:

$$V_M = \frac{100 + 1200}{100}\left[30 \times \frac{33\,(10^3)}{33\,(10^3) + 390\,(10^3)} - 0.7\right] = 21\text{V}$$

In the case of high-power amplifiers, compound configuration with split supply, as shown in Fig. 14-21, the same equations apply in somewhat different form for R_D and for the driver dissipation R_{D3} in the quiescent condition:

$$R_D = \frac{V_{cc} \times h_{FEmin}}{2\,I_{pk}} \quad \text{or} \quad \frac{R_L \times h_{FEmin}}{2}$$

In this configuration, R_D is only half the total dc load resistance that has been split and bootstrapped through C_D; I_{pk} represents the peak value of the output current; and h_{FEmin} is the current gain of the compound. That is, the product of the minimum current gains of Q5 and Q7 is given by

$$P_{D3} = \frac{V_{cc} \times I_{pk}}{h_{FEmin}} \quad \text{or} \quad \frac{2 \times P_{out}}{h_{FEmin}}$$

For the 60W amplifier, I_{pk} 3.87A; h_{FEmin} 20 x 500 1000, and V_{cc} 36V. Thus, RD may be calculated as follows:

$$R_D = \frac{36 \times 1000}{2 \times 3.87} = 4650\text{ ohms}$$

and

$$P_{D3} = \frac{36 \times 3.87}{1000} = 140\text{ mW}$$

Since R_D has been chosen equal to 3900 ohms, P_{D3} equals 170 mW.

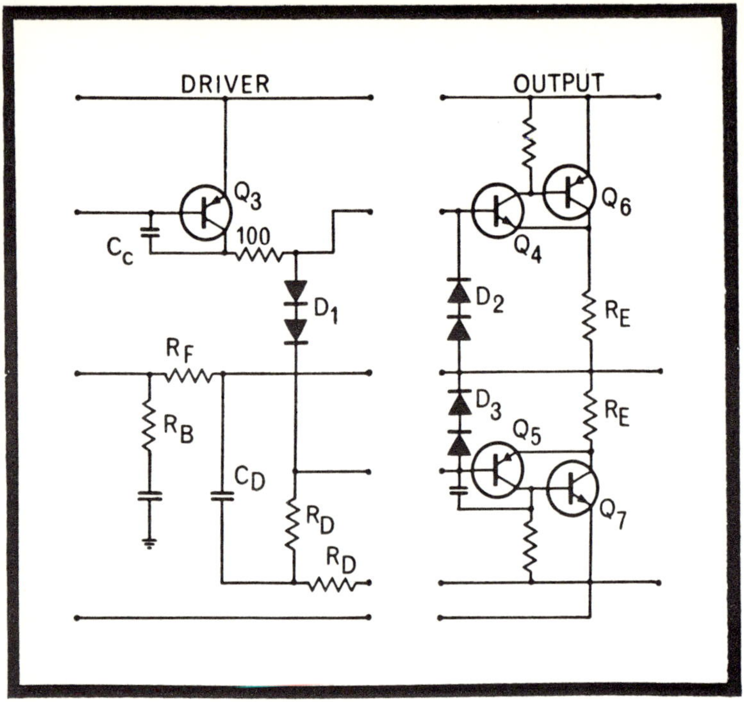

Fig. 14-21. Driver stage for high-power amplifier.

Take note that the dissipation P_{D3} of the class A operated driver is much smaller for the 60W amplifier than it is for the 10W amplifier. It also results that, for moderate powers of 10W or less, a compound configuration is recommended for low-value loads when the currents are high and the gain is small.

The **voltage gain** of Q3 is given by

$$g_3 = \frac{Vcc}{26\,mV}$$

at room temperature. The load lines for Q3 are quite similar to those of Fig. 14-20. For the 60W amplifier, the voltage gain of Q3 is equal to

$$\frac{Vcc}{26\ mV} \quad \text{or} \quad \frac{36}{26} \times 10^3 = 63\,dB$$

Because the total closed-loop ac gain is given by

$$\frac{R_F + R_B}{R_B} \quad \text{or} \quad \frac{10000 + 470}{470} = 22 = 27\,dB$$

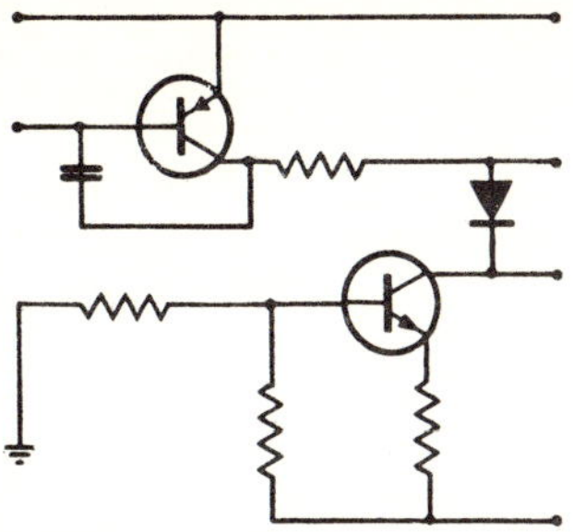

Fig. 14-22. Current source used as load for the driver.

the total feedback ratio will be 46 dB with an input-stage gain of 10 dB. The total dc voltage for optimum stability is unity. A small (100 ohm) resistor limits the current, which would otherwise reach too high a value in fault conditions. The smaller capacitor C_c between the collector and the base of Q3, multiplied by the Miller effect, sets the single dominant high-frequency pole far away from the basic turnover frequencies of the input and output circuits. In this way, the amplifier as a whole demonstrates a 6-dB-per-octave rolloff characteristic from about 50 kHz. A large amount of feedback can, therefore, be applied to decrease distortion while maintaining stability.

Resistor R_D and capacitor C_D can be replaced by a current source, as shown in Fig. 14-22. Such a system works exactly as in the former case.

INPUT STAGE

The **low-power amplifiers** from 1 to 10W described herein include an input stage working as a preamp. They are,

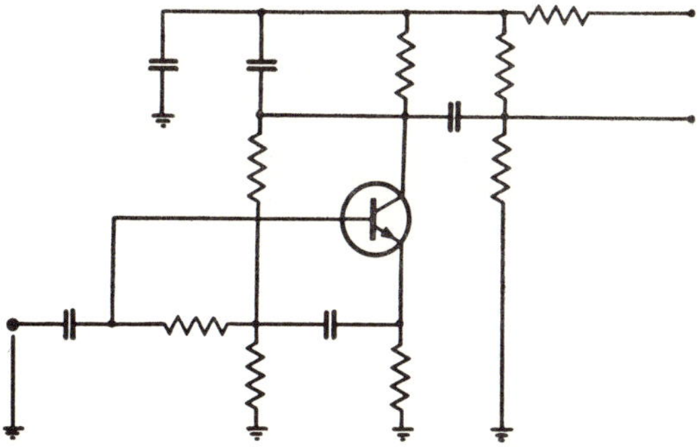

Fig. 14-23. Input stage (preamp) of the 10W amplifier.

therefore, complete amplifiers of high impedance and high sensitivity, suited for direct connection to phonograph cartridges. The function of the preamp is to provide a certain voltage gain and high input impedance.

The voltage gain in the example (Fig. 14-23) is 10. The high input impedance is obtained by bootstrapping. The collector load resistance is shunted with a small capacitor to provide a high-frequency rolloff and to stop parasitic oscillation (caused by circuit layout) that may develop between the preamp and the basic power amplifier.

All of the **high-power amplifiers** presented in this book have a sensitivity of 1V rms into a 10K load for full rated output. This choice determines their gain. The input stage has a gain of only 10 dB, and normally consists of a differential amplifier (dif-amp), as in Fig. 14-24.

The dif-amp scheme is used because of the inherently temperature-compensated balancing of both base legs. Controlled symmetry of voltages in these legs is the simplest technique for providing a desired remote voltage level (the amplifier output).

Transistor Q1 is biased through a resistor in its base lead to ground. Transistor Q2, biased through a similar resistor, will cause the output to be equivalent ground because of the symmetry of the two halves of the local circuit. The error voltage, appearing as an offset at the output, consists of the differential base-to-emitter voltage between the two transistors and the differential base current flowing through the

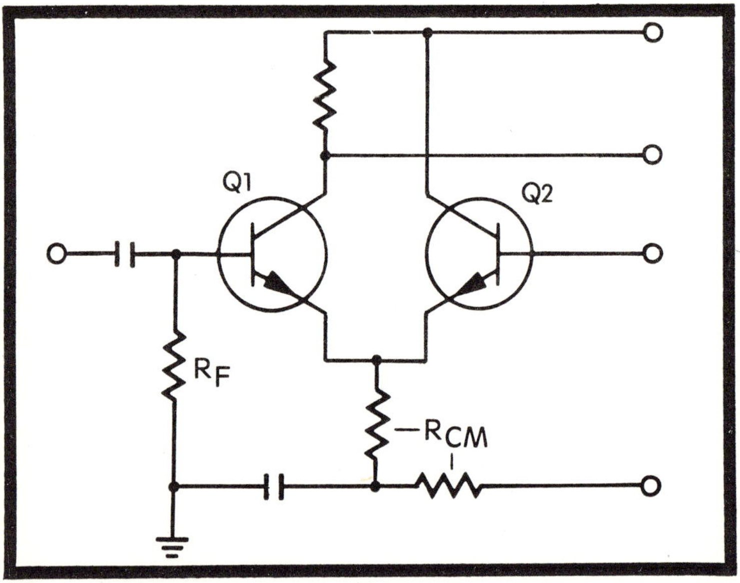

Fig. 14-24. Input circuit differential amplifier.

bias resistor (R_F). When Q1 and Q2 form a dual transistor in one package, typical offsets of less than 10 mV are realized at the output.

Transistor Q1 also serves as a common emitter to the input signal, and uses the common-base input impedance of Q2 as its effective emitter load. Constant-current emitter dc bias for Q1 and Q2 is supplied through the common-mode path (R_{cm}), and decoupling is provided in this path for optimum power-supply ripple rejection.

POWER MEASUREMENT

Reference is made here to the "IHF Standard Methods of Measurement for Audio Amplifiers" or standard IHF-A-201, published by the Institute of High Fidelity, Inc. This standard makes a distinction between the **continuous output**, as "the largest single-frequency output obtainable with a sine wave signal for a substantial length of time at reference distortion" and the **dynamic output**, as "the largest output obtainable with a sine wave signal for a short period of time at reference distortion."

The continuous output, more often called the rms output, depends upon the supply used. The dynamic output, also called music power of IHFM power, is measured "with all significant voltages maintained at the same value (by means of external supplies for example) as they were under no-signal conditions." All of the amplifiers described in this book are rated at a certain amount of music power, which can be equal to the rms power if the supply is substantial and regulated.

PROTECTION CIRCUITS

In power amplifiers, the output devices are always used near their limits; in certain conditions of operation, these limits can be exceeded. Consequently, a protection circuit should be able to sense current, voltage, and power, and to limit each of these quantities as soon as any of these quantities exceeds the SOA.

An integral protection, therefore, should guarantee first a current limitation—3 to 5A, depending upon duration for the MJE371; 30 to 50A for the MJ802; then a voltage limitation determined by BV_{CE} under certain bias conditions; and finally, a power limitation because of the maximum allowed dissipation P_D (temperature dependent) or because of second breakdown (not temperature dependent).

A current or a voltage limitation can be made relatively simply. In practice, many amplifiers are only current limited; however, such a solution is generally dictated by economical reasons, and merely constitutes a compromise. A more useful approach is to make a power limitation.

To any power pulse P_p of any duration, repetitive or not, there will correspond a final junction temperature Tj that will

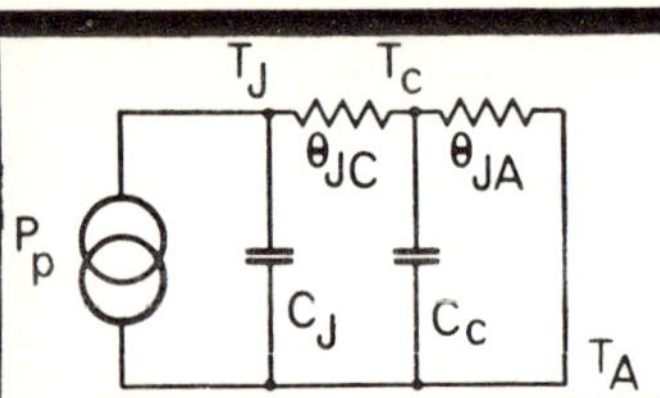

Fig. 14-25. Relation between power pulse and junction temperature.

be established through a system of time constants (Fig. 1-25). The protection circuit must be an electrical image of the above thermal relations. The P_p must be converted into a current I that will generate a voltage V_j proportionally related to T_j through an appropriate RC network (Fig. 14-26). As soon as T_j reaches T_{jmax}, V_j will be V_{jmax}, and this voltage must then be used to switch the device off instantaneously or to decrease the amplifier gain. Because two different kinds of power limitation exist, two different considerations must be made.

For protection against excessive power limitation P_D, the pulse of power P_p will be equal to the voltage V_{CE} multiplied by the collector current Ic of the device to be protected. To build Pp requires the true multiplication of two electrical parameters. The pulse may not have a rectangular shape, but this assumption is made for the sake of simplicity. The RC network can be derived from the curve of the transient thermal resistance vs time.

This limitation further depends upon the temperature T_A. A second voltage V_A proportional to T_A must, therefore, be added to the RC network. The resulting voltage V_j will be used to decrease the gain (Fig. 14-26). To include the response time of the protection circuit until the gain is reduced, the total time between pulse beginning and gain reduction must equal the maximum allowable pulse duration, as indicated by the SOA.

For protection against second breakdown, the power limit to be considered will be proportional to the multiplication of V_{CE} and I_c where V_{CE} will be affected by an exponent larger than 1 (e.g., 2) corresponding to the steeper portion of the curve.

In this case, the double product $V_{CE} \times V_{CE} \times I_c$ must be formed. Furthermore, when the pulse duration decreases, the

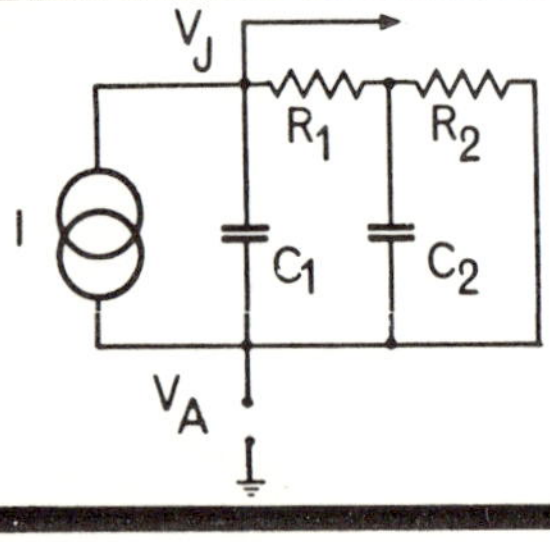

Fig. 19-1. The vertical sweep circuits drive the vertical yoke and vertical convergence coils. (Courtesy Sylvania.)

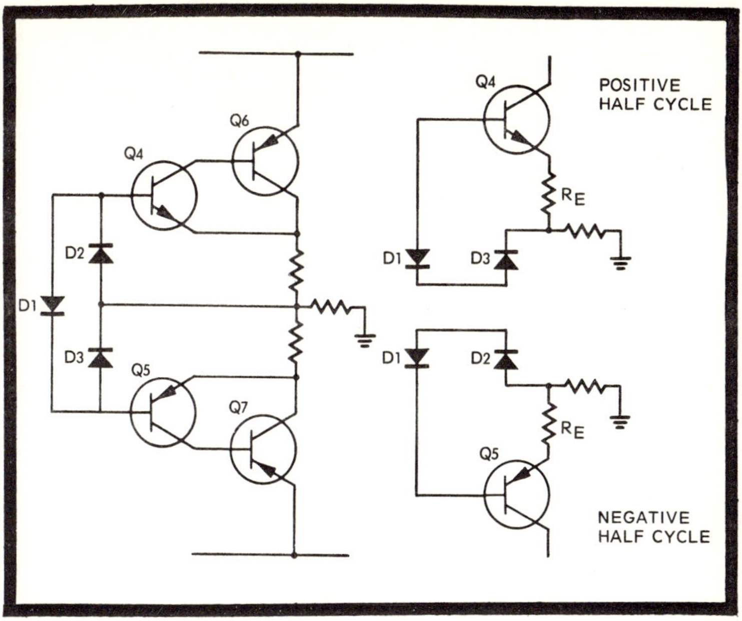

Fig. 14-27. Overload protection using diodes.

exponent of V_{CE} changes, thus creating a still more complex situation. As a compromise, a circuit similar to the former one can be used. It will differ by the input current obtained from conversion of P_p, which will vary according to another law as a function of voltage and current. It will also differ from the RC network because of the reduction of the active surface of the die and the absence of temperature dependence. In the text following, some practical circuits, working efficiently within certain limits, will be discussed.

In the case of power amplifiers for audio applications, the limits to be considered are from very low to very high frequencies. Because safe operation is more critical at low frequencies, however, only this case will be investigated; the RC network discussed above need not be used in the protection circuit. Furthermore, nonsaturated and saturated conditions must be considered because large differences in dissipation may be encountered.

Simulation of the actual signals by sine waves is a simplification. In fact, signals may have any shape. The longer they are, the more severe are the conditions for the devices. Among all of the possible signals those generator accidentally (by rubbing the pickup needle, for instance!) should also be taken into account.

Current limitation by the use of diodes (Fig. 14-27) shows the principle of this type of protection. It is employed for some

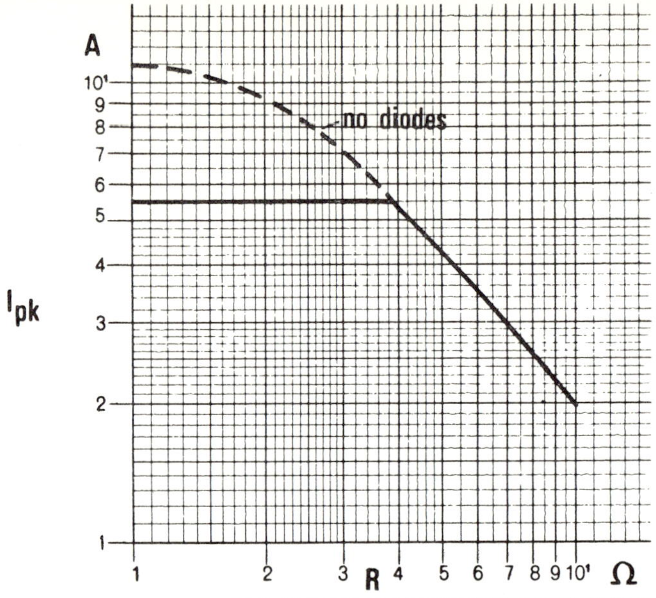

Fig. 14-28. Peak current as a function of load resistance at 1 kHz for the MJ2841 of the 60W amplifier.

high-power amplifiers (shown later), but is readily extendable to lower powers. To provide load-fault protection, diodes D2 and D3 work in conjunction with D1. On the positive half cycle of the output signal, D1 and D3 will appear in series between the base of Q4 and the output, as shown in the figure, and will limit the peak-fault current to a chosen value (through R_E) slightly above the normal peak-load current by limiting the voltage between the base of Q4 and the output.

On the negative half cycle, fault protection is provided by D4 and D2. For the 60W amplifier, the current-limiting action is shown in Fig. 14-28 as a function of the load resistance. Considering the short-circuit condition, it is found that the device is then submitted simultaneously to high voltage and high current values—33.8V and 5.2A, respectively, as shown in Fig. 14-29.

The peak power is 33.8 x 5.2 equals 175W. The SOA of the MJ28411 shows that the second-breakdown limitation imposes a maximum duration of 3 msec for such a pulse, independently of the case temperature. A limitation by the power dissipation for a single pulse of the same length will only occur at a case temperature of

$$T_C = T_j - P_p \times r \times \theta_{jc},$$

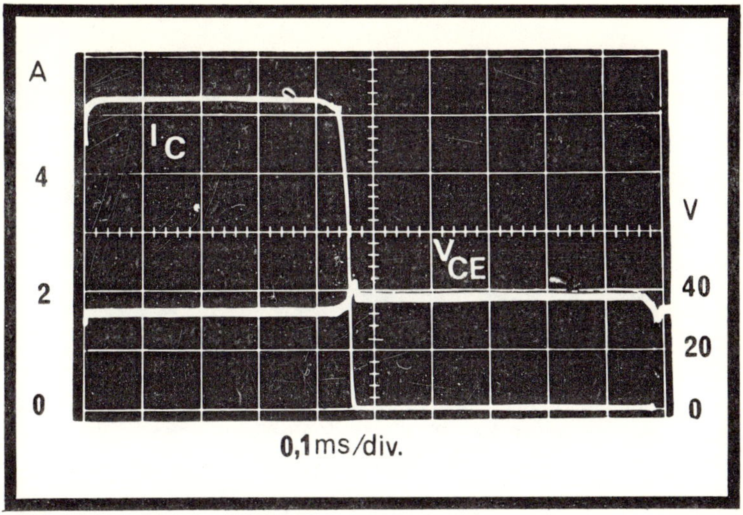

Fig. 14-29. Short-circuit test for the 60W amplifier at 1 kHz. Collector current and collector-to-emitter voltage vs time for the MJ2841.

where r is the normalized transient thermal resisance, or 200 —175 x 0.38 x 1.17=122 deg C or more.

If a feedback loop is used to decrease the gain or the input signal for total protection, it should have a response time of 3 msec.

The above short-circuit current can be calculated from

$$\frac{V_{D1} + V_{D2} - V_{be4}}{Re}$$

Typically, its value is equal to

$$\frac{1.35 + 1.35 - 0.65}{0.43} = 4.8A$$

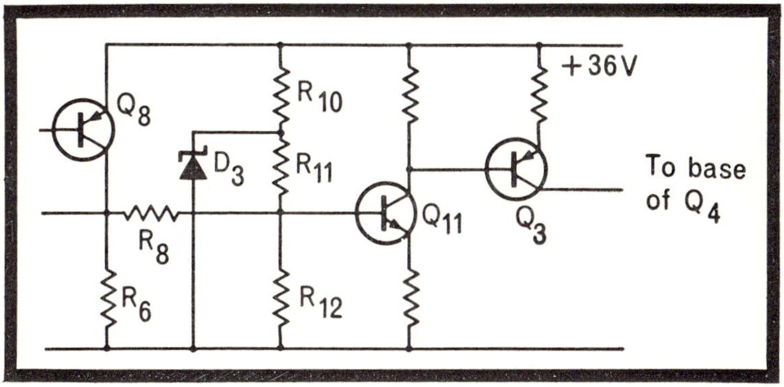

Fig. 14-30. Protection circuit sensing current and voltage.

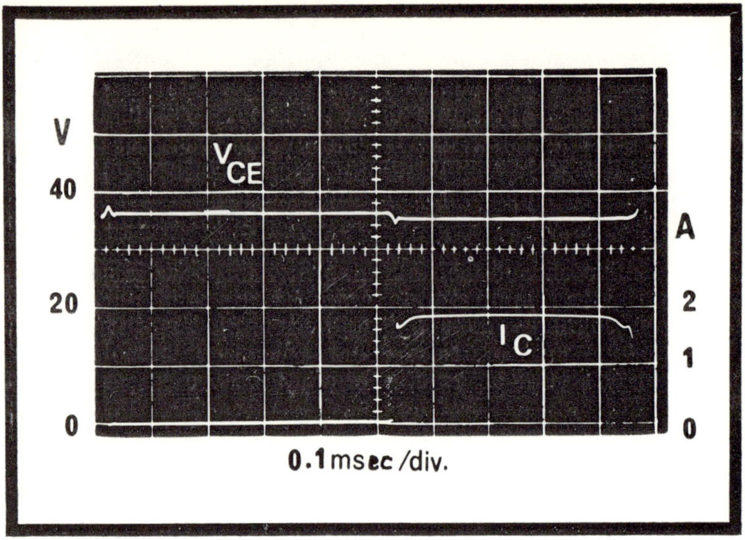

Fig. 14-31. The V_{ce} and I_c for short-circuit condition.

The voltages V_{D1}, V_{D2}, and V_{D3} have a tolerance of plus or minus 4 percent. Resistor R_E is a plus or minus 5 percent tolerance resistor. This type of protection is only current dependent. If the load is reactive, V_{ce} will be higher than 36V and, at the same time, a collector current will flow. For large phase shifts and low impedances, simultaneous high-current and high-voltage values are possible. This case is examined for the protection circuit described below.

Protection by voltage-dependent current limitation is shown in Fig. 14-30. It makes use of the superposition of a current proportional to the collector current I_c of Q8 and a current proportional to the voltage V_{ce}, realized in the summing network R8-R10-R11-R12. With the 60W amplifier, the limitation is determined by the current circuit only for voltages above 28V. Below that value, the voltage circuit comes into play by decreasing the total current through Q11. The limited value of the output current is thus increased as the output voltage is decreased. Protection is obtained by turning Q3 on, thus turning Q4 off.

In short-circuit conditions, the voltage V_{ce} is nearly equal to V_{CC} and the current will be limited to 1.8A as shown in Fig. 14-31.

The MF2841 and MJ2941 output devices can withstand this dissipation indefinitely, as long as the case temperature T_c does not exceed $T_c = T_j - \theta_{jc} \times P_c$, or 200 — 1.17 x 1.8 x 36=124 deg C.

The total V_{ce} — I_c limited curve of this protection has the shape indicated in Fig. 14-32. This curve can be obtained by measurement of the maximum values of V_{ce} and I_c when the load impedance and phase are varied. A comparison with the corresponding SOA, considering very low frequencies, shows that operation may not be safe in certain regions of simultaneously high current and high voltage. Peak and average powers will be calculated below as a function of the load phase before saturation.

The situation in conditions of saturation—with an input voltage of 1V, a load of 8 ohms, and a 60 deg phase angle—is shown in Fig. 14-33 for an example. The limiting curve can also be determined from this figure. Starting at 28V, the maximum current can be seen to increase from 1.8A to about 4.5A at zero volts V_{ce}.

Figure 14-34 shows the power curve vs time, as determined from Fig. 14-33 by multiplication of V_{ce} by I_c. The peak power is 51V x 1.8A or 92W, and has an equivalent width 0.25t, where t is the period. To determine whether or not the devices are operated safely, refer to SOA discussed above.

Larger phase shifts would lead to still worse conditions. However, a phase shift or 60 deg is generally the maximum value encountered with loudspeakers. It can be found on both sides of speaker resonance, and has inductive and capacitive characteristics depending upon the side of resonance measured. The impedance and phase-angle curves of the speaker used should always be known, and tests should be made in worst-case conditions, especially at maximum phase shift.

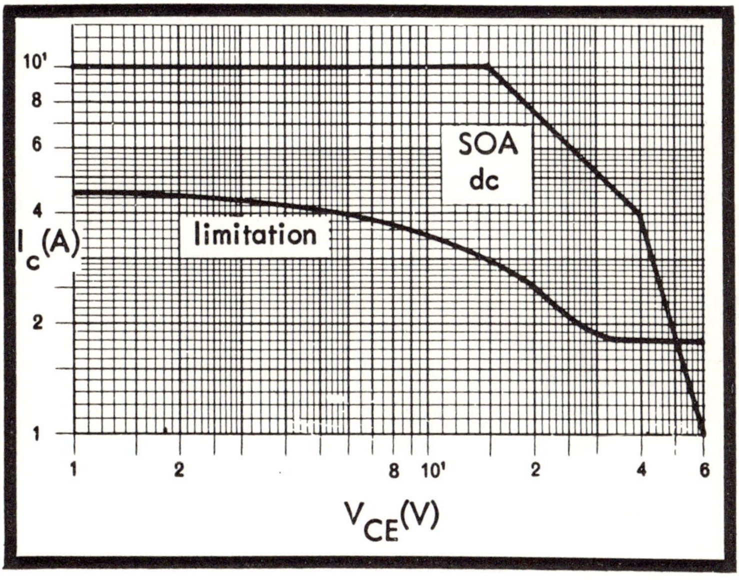

Fig. 14-32. Limitation curve compared to SOA for dc conditions.

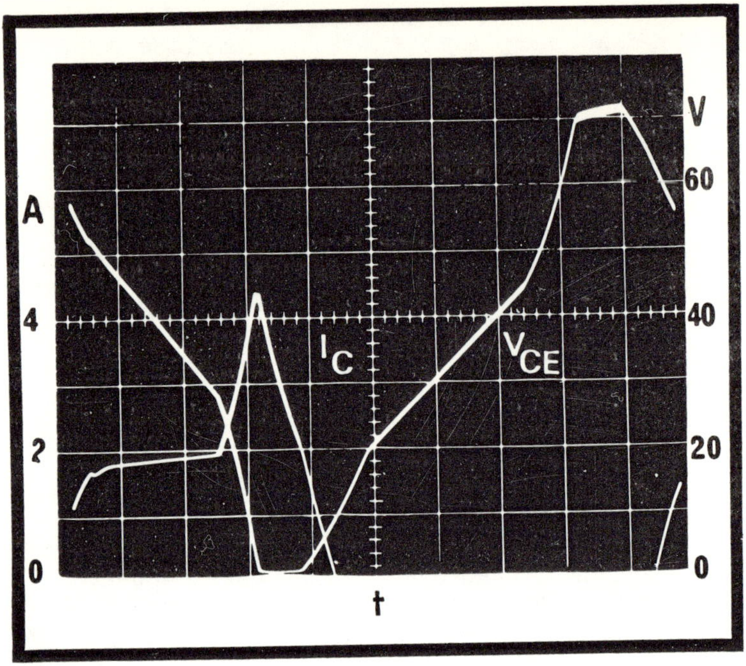

Fig. 14-33. The V_{ce} and I_c vs time for an 8 ohm, 60 deg phase-angle load.

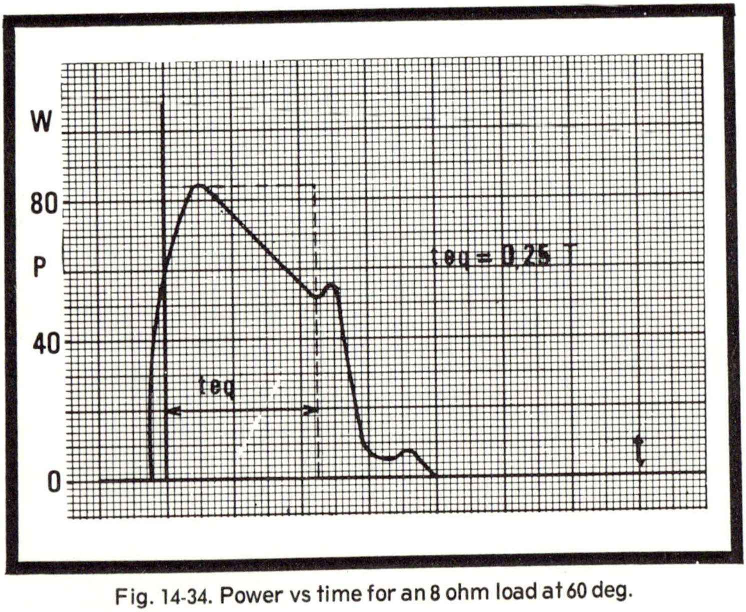

Fig. 14-34. Power vs time for an 8 ohm load at 60 deg.

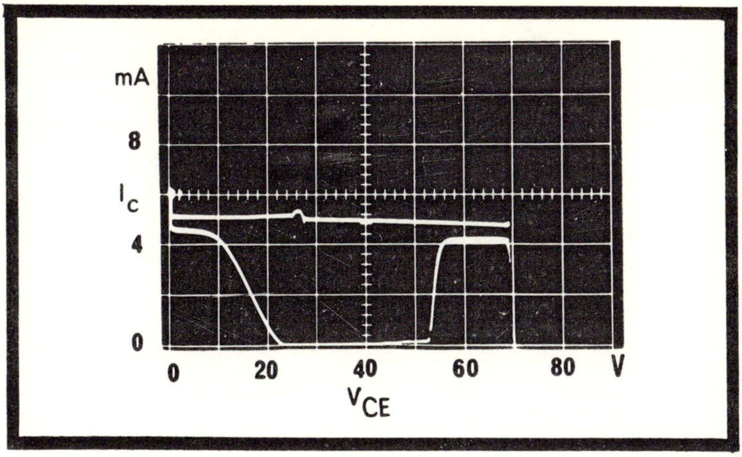

Fig. 14-35. Load line for the driver with an 8 ohm output load at 60 deg.

Figure 14-35 shows the load line for the driver of the 60W amplifier in the same conditions of input drive and output load as those of Fig. 14-33, with the protection circuit connected.

THERMAL STABILITY

Refer again to Fig. 14-5, which shows the variation of the quiescent current I_Q as a function of the ambient temperature T_A for the 60W amplifier. The measurement has been made at two different power levels, P_{D1} and P_{D2}, corresponding respectively to minimum and maximum dissipation. Note that the current increase in both cases is very small. Investigation of the thermal stability under dynamic conditions can be made when passing abruptly from P_{D1} to P_{D2}. This test furnishes an experimental method of determining the most convenient mounting place for the bias diodes.

Figure 14-36 shows the resulting quiescent current as a function of time, beginning immediately after interruption of the input signal. In this configuration, the diodes match the drivers; therefore, when diodes and drivers are mounted on the printed circuit (PC) board, the transient period is negligible. Conversely, when the diodes are mounted on the same heat sink as are the output transistors, a transition time of several minutes results, with complete cancellation of the quiescent current through the output devices. However, both situations are stable. Thermal runaway may result if the diodes cool down more rapidly than do the drivers.

RF STABILITY

Additional compensation networks are used at the output of the high-power amplifiers. Among these are an RC pad

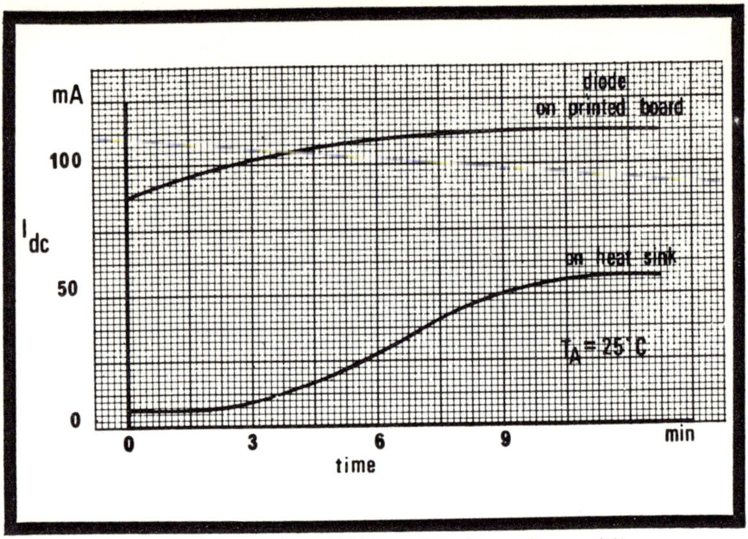

Fig. 14-36. Thermal stability under dynamic conditions.

(inserting a stabilizing pole and zero into the amplifier Bode response) across the output and a choke in series with the load. The latter is a low-resistance rf-type choke used to provide a buffer between the amplifier and a capacitive load. With many amplifiers, across the output will upset stability and cause oscillation because it provides an uncompensated pole in the Bode response. A small inductance (about 2 uH) is used to eliminate this effect.

It is also advisable to add power-supply bypass capacitors near or on the PC board. These capacitors are used to provide very-high-frequency return paths to the amplifier. The high-load currents flowing through long power-supply lead wires (consisting of series R and L) generate enough ground loop at very high frequencies to cause the amplifier to oscillate. Ceramic-disc bypass capacitors are required to eliminate this source of parasitic oscillation. It might also be necessary to add a 0.001 uF capacitor directly across the bases of Q6 and Q7.

THEORETICAL DESIGN PROCEUDRES

For peak and average values of dissipated power for each transistor of a complementary pair when used in class B:

$$\frac{VCE}{Vo} = 1 + M1 \times \cos(\omega t)$$

$$\frac{I_C}{I_O} = M2 \times \cos(\omega t -)$$

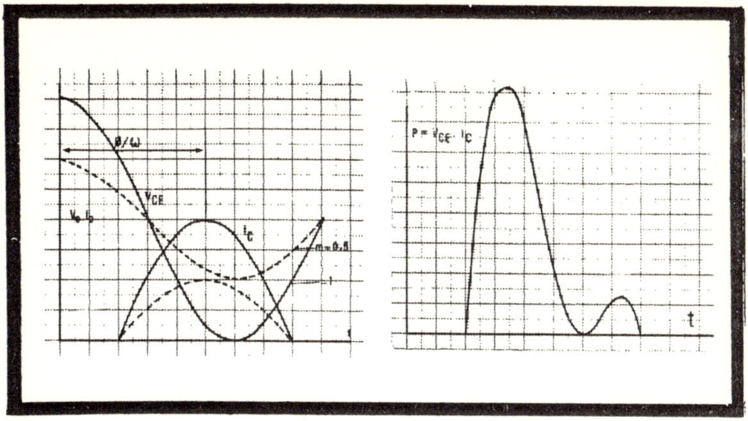

Fig. 14-37. Voltage, V_{ce}, current I_C, and instantaneous power P vs time.

where θ is the phase difference between V_{ce} and I_c,

$$m_1 = \frac{V_{ce}}{V_{cemax}} \quad \text{and} \quad m_2 = \frac{I_c}{I_{cmax}}$$

m_1 and m_2 may vary between 0 and 1.

With $P = V_{ce} \times I_c$ and $P_o = V_o \times I_o$, normalized instantaneous power is equal to

$$\frac{P}{P_o} = (1 \times m_1 \cos(\omega t) \times m_2 \cos(\omega t - \phi)$$

$$= m_2 \times \cos(\omega t - \phi) + \frac{m_1 \times m_2}{2} \cos\phi + \frac{m_1 \times m_2}{2} \cos(2\omega t - \phi)$$

At peak values of power, the derivative of P / P_o is equal to zero. When $m_1 = m_2 = 1$, it becomes

$$\sin(\omega t - \phi) + \sin(2\omega t - \phi) = 0$$

After the above equation has been solved, two peaks are found at t_1 and t_2, respectively:

$$t_1 = \frac{2}{3\omega} \quad \text{or} \quad \frac{t_1}{t_o} = \frac{\phi}{3\pi}$$

$$t_2 = \frac{2}{3\omega}(\pi + \phi) \quad \text{on} \quad \frac{t_2}{t_o} = \frac{\pi + \phi}{3}$$

where t_o is the period. See Fig. 14-37.

The normalized peak powers are equal to

$$\left(\frac{P}{P_o}\right)_1 = \frac{1}{2}\cos\phi + \frac{3}{2}\cos\frac{\phi}{3} \text{ at } \frac{t_1}{t_o}$$

$$\left(\frac{P}{P_o}\right)_2 = \frac{1}{2}\cos\phi + \frac{3}{2}\cos\left(\frac{\phi}{3} - \frac{2\pi}{3}\right) \text{ at } \frac{t_2}{t_o}$$

Note that varies between (resistive load) and $\pi \pm \pi/2$ (purely reactive load). Figure 14-38 shows both functions $(P-P_o)_1$ and $(P-P_o)_2$ vs ϕ. Figure 14-39 gives the normalized times t_1/t_o and t_2/t_o at which peak powers occur vs ϕ. Also, see Fig. 14-40.

The worst-case conditions occur for $\phi = \pi/2$. The normalized peak powers are then equal to 1.3 and zero.

For the 60W amplifier, this situation would correspond to the following numerical values:

$V_o = 36V \quad I_o = 3.9A$

$P_{pk} = 36 \times 3.9 \times 1.3 = 182W$

at $V_{ce} = 1.5 \times V_o = 54V$

and $I_c = 0.87 \times I_o = 3.4A$

For the output transistors used in the 60W amplifier, the MJ2841 and MJ2941, **these operating conditions are beyond the permissible limits.**

For $\phi = \pi$ (Fig.14-41), the normalized peak power is equal to

$$\frac{P}{P_o} = -m\cos(wt) - \frac{m^2}{2}\left[1 + \cos(2\omega t)\right]$$

Then,

$$\left(\frac{P}{P_o}\right)_{1 \text{ or } 2} = \frac{1}{4} \quad \text{for: } 0.05 \le m \le 1 \text{ (two peaks per cycle),}$$

$$\left(\frac{P}{P_o}\right)_{1 \text{ or } 2} = m(1-m) \quad \text{for: } 0 \le m \le 0.5 \text{ (one peak per cycle)}$$

The normalized times corresponding to the peak powers are given by

$$\frac{t}{T} = \frac{1}{2\pi} = \arccos\frac{-1}{2m} \quad \text{for: } 0.5 \le m \le 1$$

and are plotted vs ϕ, as shown in Fig. 14-41.

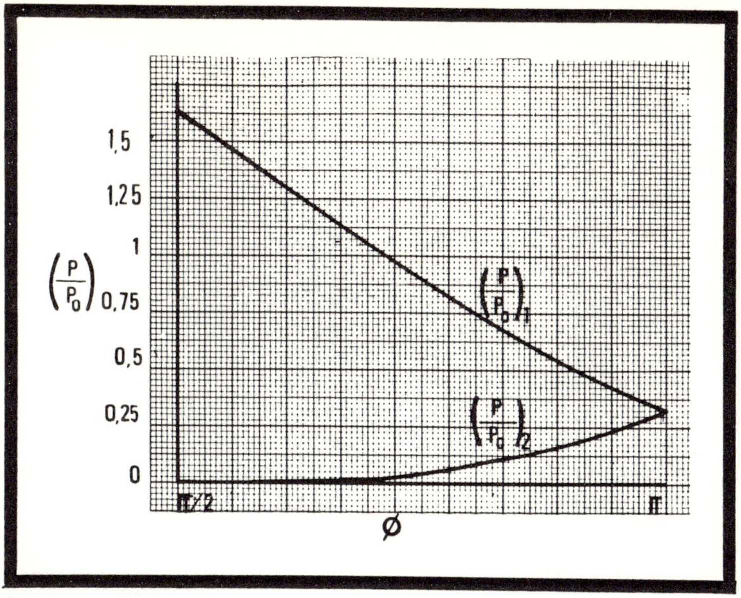

Fig. 14-38. Peak values of normalized dissipated power vs phase difference between V_{CE} and I_C.

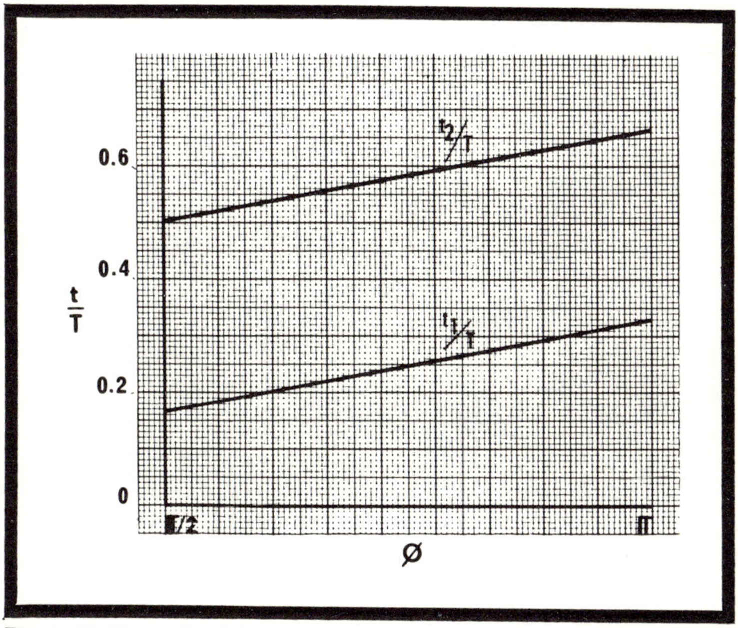

Fig. 14-39. Normalized dissipated power vs normalized time for different valves of m.

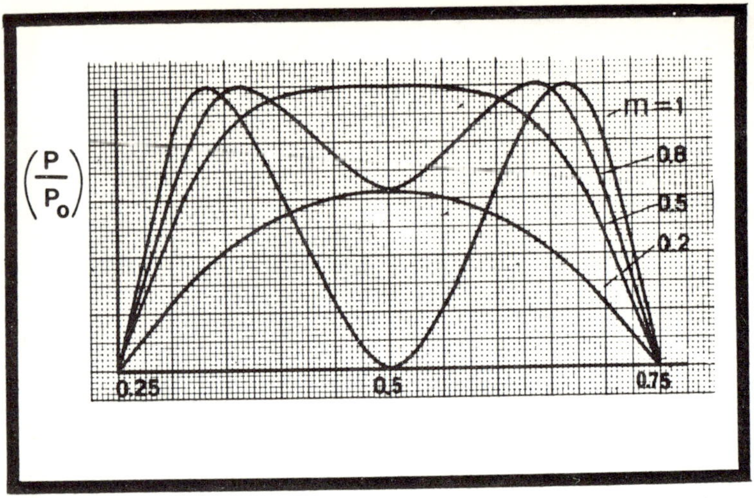

Fig. 14-40. Normalized average power vs m for different values of ϕ.

Driver Gains

The voltage gains of Q2 and Q3 in Fig. 14-43 can be approximated as follows:
(a) Voltage gain of Q3:

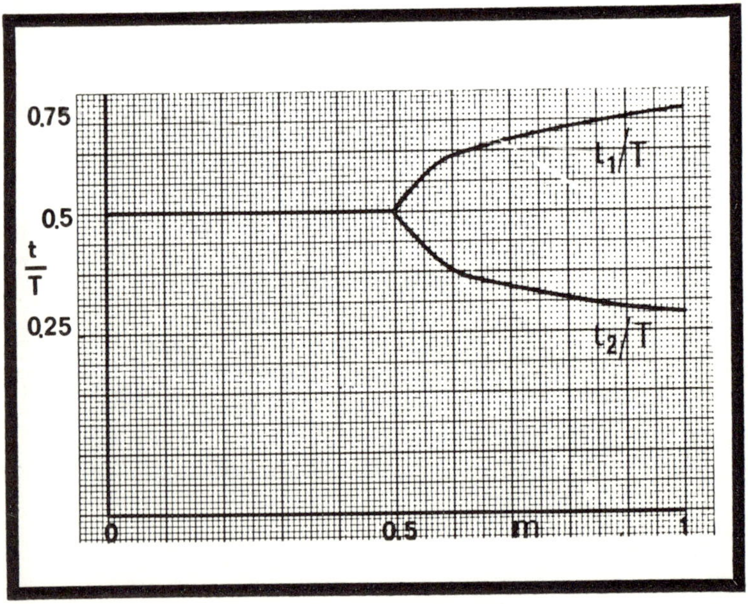

Fig. 14-41. Normalized times vs m for peak powers when $\phi = \pi$

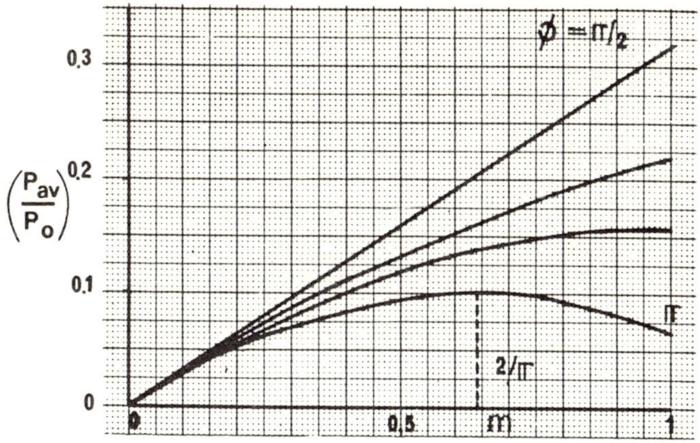

Fig. 14-42. Normalized average power vs m for different values of ϕ.

$$\frac{V_3}{V_2} = gm \times R = \frac{h_{fe-3}}{h_{11-3}} \times R$$

$$h_{11-3} = h_{fe-3} = \frac{x\,2\,G\,mV}{I_{C3}}$$

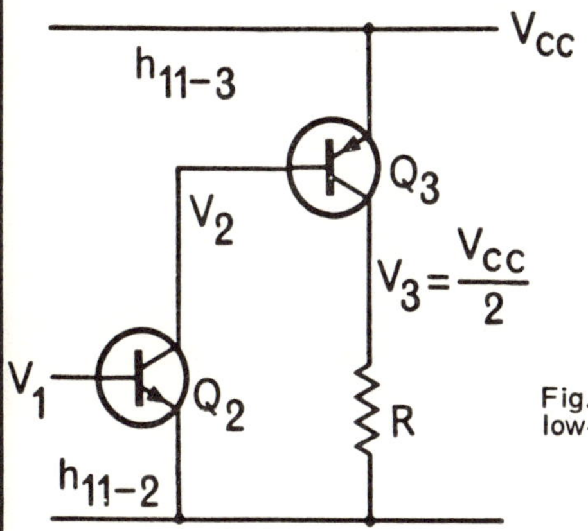

Fig. 14-43. Driver stage for low-power amplifiers.

$$\frac{V3}{V2} = \frac{Ic3 \times R}{26\,mV} = \frac{Vcc}{2 \times 26\,mV}$$

(b) Voltage gain of Q2:

$$\frac{V2}{V1} = gm \times h_{11}-3 = \frac{h_{fe}-2}{h_{11}-2} \times h_{11}-3$$

$$h_{11}-3 = \frac{26\,mV}{I_{c1}} \quad h_{11}-2 = h_{fe}-2 \frac{\times 26\,mV}{I_{c2}}$$

After the above equations have been solved, it appears that the gain of Q2 is equal t $V_2 / V_1 = 1$.

Load Impedance of Driver when Bootstrapping

Figure 14-44 shows the situation for the positive swing in the noncompound configuration. Neglecting small voltage drops, such as V_{be}, the load seen by Q3 is equal to

$$RQ3 = \frac{Vcc/2 + Vo}{Itot}$$

with $m = \frac{Vo}{Vcc/2}$ and $K = \frac{hFE}{hFEmin}$

and the normalized value of RQ3 will be:

$$\frac{RQ3}{hFEmin \times R_L} = \frac{1+m}{1+m/k}$$

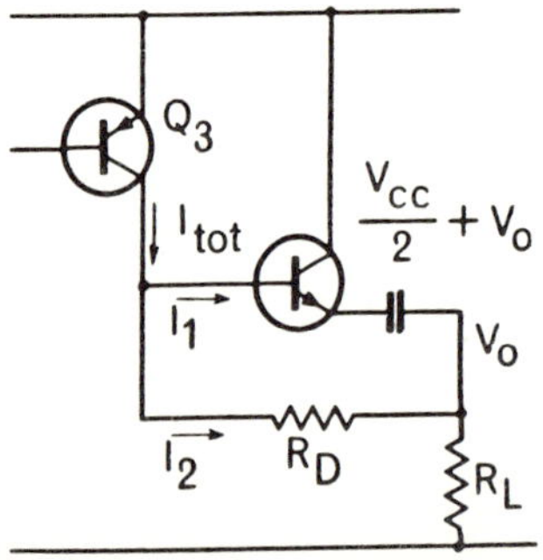

Fig. 14-44. Load impedance of Q3 for positive swing $R_D = R_L \times h_{FEmin}$.

The normalized collector-emitter voltage of Q3 is equal to

$$\frac{V_{Q3}}{V_{cc}/2} = 1 - m$$

and its normalized collector current:

$$\frac{I_{Q3}}{I_o} = 1 + \frac{m}{k}$$

Figure 14-45 shows a normalized load line as a function of the output swing for different values of $h_{FE} - h_{FEmin}$.

THERMAL STABILITY UNDER DYNAMIC CONDITIONS

Figure 14-46 shows the basic circuits used for this investigation. The criterion for thermal stability under dynamic conditions is a decrease of dissipated power when the transistor passes abruptly from a certain dissipation level to one that is normally lower. During this level change, the junction temperature of the transistor should remain constant while the bias diode may cool down rapidly, giving rise to a bias voltage increased.

$$P_{D1} = V_{ce} \times I_{c1}$$

$$P_{D2} = V_{ce} \times I_{c2} \quad \text{at same } T_j$$

$V_{BE} \lhd K \times \rhd T_j$, where $K = 2$ mV per deg C;
$I_{c2} = I_Q + \rhd I_c$, where $\rhd I_c$ is the current increment over I_Q during the transition period, and is equal to

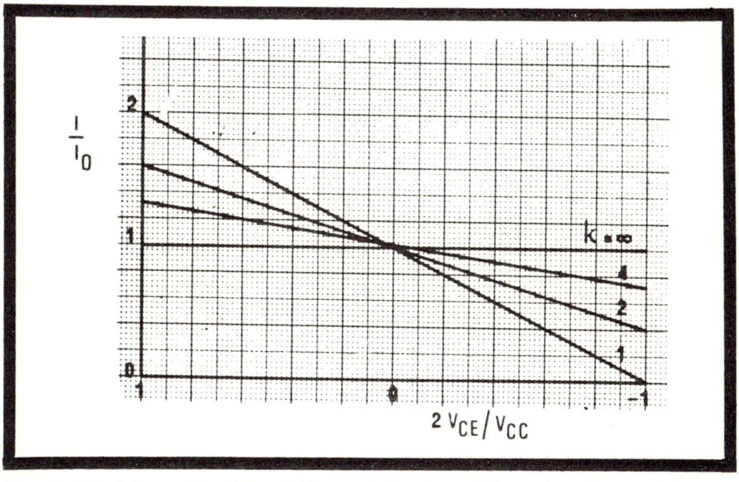

Fig. 14-45. Normalized load lines of Q3 for different values of $h_{FE} - h_{FEmin}$.

$$I_c = \frac{\Delta V_{be}}{R_E} = \frac{K \times \Delta T_j}{R_E} = \frac{PD1 \times \theta_{JA} \times K}{R}$$

Solving the above equation gives the condition for stability:

$$P_{D1} > V_{ce} \times I_Q + \frac{P_{D1} \times \theta_{jA} \times K}{R_E}$$

or the condition for R_E:

$$R_E > \frac{P_{D1} \times V_{CE} \times \theta_{jA} \times K}{P_{D1} - V_{CE} \times I_Q}$$

For the driver stability in the compound configuration (Fig. 14-46b), one must consider R_E (h_{FEmin} + 1), where h_{FEmin} is the minimum current gain of the output device. Then the condition for R_E, for negligible I_Q, becomes

$$R_E > \frac{V_{ce} \times \theta_{jA} \times K}{h_{FEmin} + 1}$$

With the 60W amplifier, the values are V_{cc}=36V andθ_{JA}= 175 deg C per watt (thermal resistance of driver without heat sink); h_{FEmin} is equal to 20 at 4A. Thus, R_E must be greater than

$$R_E > \frac{36 \times 175 \times 2 \times 10^{-3}}{21} = 0.6 \text{ ohms}$$

A heat sink on the drivers allows this value to be reduced.

In the **compound** circuit, the output transistor current is controlled by the driver. However, because R_E is common to the driver and output transistor, any change in emitter-base voltage of the power device is counteracted by a change of the

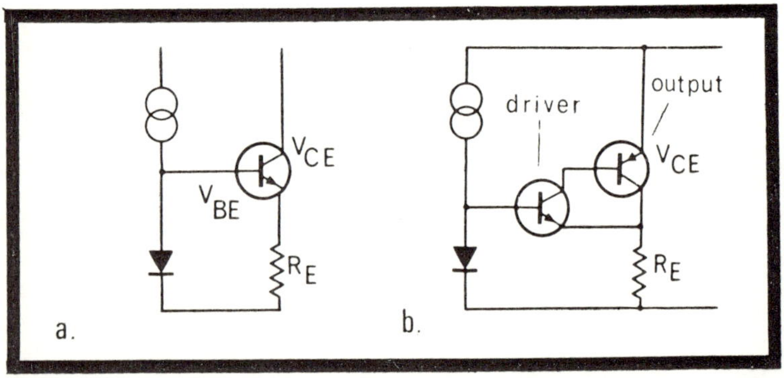

Fig. 14-46. Basic circuit for thermal stability.

voltage at R_E multiplied by the gain of the driver (slightly less than 1).

When applied to the output transistor, the condition for stability gives the following results: $R_E = 36 \times 4.47 \times 2 \times 10^{-3} = 0.32$ ohms, where 4.47 is the sum of the thermal resistances of the output transistor (1.17 deg C per watt), and heat sink (3 deg C per watt). However, it is not necessary to fulfill this condition.

RECTIFIER

The calculation that follows is based upon the schematic and the simplified diagram shown in Fig. 14-47. Full-wave rectification will be considered.

Given parameters: E_{dc} and I_{dc}. Parameters to be calculated: E_S, I_D, I_{TR}, and C.

The total conduction angle is 2ϕ:

$$\cos \phi = \frac{E_{DC} + 2E_D}{E_{pk}}$$

The total resistance of the rectifier circuit is given by

$$Rt = \eta^2 \times R_p + R_s + 2 R_D,$$

where R_p and R_s are the primary and secondary winding resistances, respectively. The dynamic resistance of one rectifier is represented by R_D.

The efficiency of the circuit is equal to:

$$\eta = \left(\frac{R_L}{R_L + Rt}\right)^2$$

Here is the suggested method for the calculation of the parameters using certain assumption:

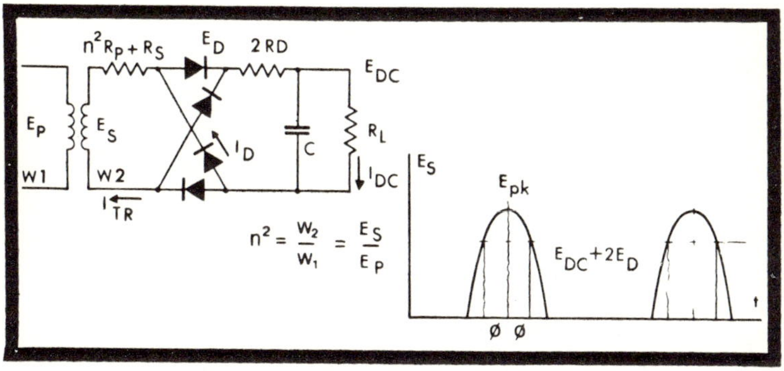

Fig. 14-47. Basic circuit schematic.

RL/Rt	f2 (ohms or microfarads)
5.	10.0
6.5	15.0
8.	20.5
10.	27.0
10.	32.0
15.	38.0
20.	50.0
30.	65.0
50.	80.0
100.	100.0
200.	130.0
300.	145.0
500.	165.0
1000.	200.0

Table 14-2

(1) Assume a value for η (e.g., 85 percent). It follows that $R_t = 0.085\ R_L$.
(2) Take E_S(rms) E_{DC} and determine the function:

$$\frac{Idc \times \pi \times Rt}{2\ E_{pk}}$$

This is equal to sin ϕ - ϕcos ϕ, and is called f_0 in Table 14-1:

$$E_{pk} = \sqrt{2 \times I}$$

(3) Using Table 14-1, determine f1:

$$f_1 = \cos\phi = \frac{Edc + 2E_D}{E_{pk}}$$

$E_D = 0.7V$ per diode.
(4) Using E_{pk}, E_D, and f_0, calculate E_{dc}:

$$E_{dc} = f_1 \times E_{pk} - 2\,E_D$$

This value must be equal to the given value of E_{DC}. If not, assumption (2) is not accurate enough, and E_S(rms) cannot be made equal to E_{dc}. Change it until agreement is found.
(5) To determine C, form the ratio R_L / R_t and use the function f2 in Table 14-2; f_2 is based upon the variation of rectification efficiency E_{dc}—E_{pk} vs shunt capacitance C. The

value of f_2 given in Table 14-2 has been determined so that an increase of f_2 would not improve the efficiency: $C = f_2 / R_L$ in microfarads.

(6) Determine the peak diode current by use of f_3 in Table 14-1:

$$I_{D(pk)} = \frac{\pi(1 - \cos\phi)}{2 f_o} \times I_{dc} = f_3 \times I_{dc}$$

(7) Determine the rms value of the current passing through each diode, using f_4 in Table 14-1:

$$I_{D(rms)} = \frac{\sqrt{\pi} \times \sqrt{2\phi - 3/2 \times \sin 2\phi + \phi \cos \phi}}{2\sqrt{2}\,(\sin\phi - \phi\cos\phi)}$$

It is also important to know the current passing through the secondary winding of the transformer because it determines its dissipation. This current is equal to

$$I_{TR(rms)} = \sqrt{2} \times f_4 \times I_{dc}$$

(8) Determine the surge current through the diodes (during the switch-on period) by $I_{D(surge)} = E_{pk} / C_{rt}$.

(9) The peak value of the ripple voltage is given by

$$V_{r(pk)} = \pm\, 1\, 2 \times \frac{T_D}{f_2} \times E_{dc}$$

$$\text{where } T_D = \frac{(1 - 2\,\phi)}{\pi}$$

T in seconds. For a line frequency of 60 Hz, T=0.0167 sec. The dc current delivered to and absorbed by the described amplifiers has the form shown in Fig. 14-48.

The calculation of the rectifier has been carried out for a constant load current I_{dc}. However, because the actual current delivered to the amplifier is a half-wave rectified current at the signal frequency, a correlation must be established between this current and an equivalent dc current.

$$I_{dc} = \frac{1}{\pi} I_{pk} = \frac{V_{cc}}{2 R_L}$$

for single supply, or

$$\frac{V_{cc}}{R_L}$$

for split supply.

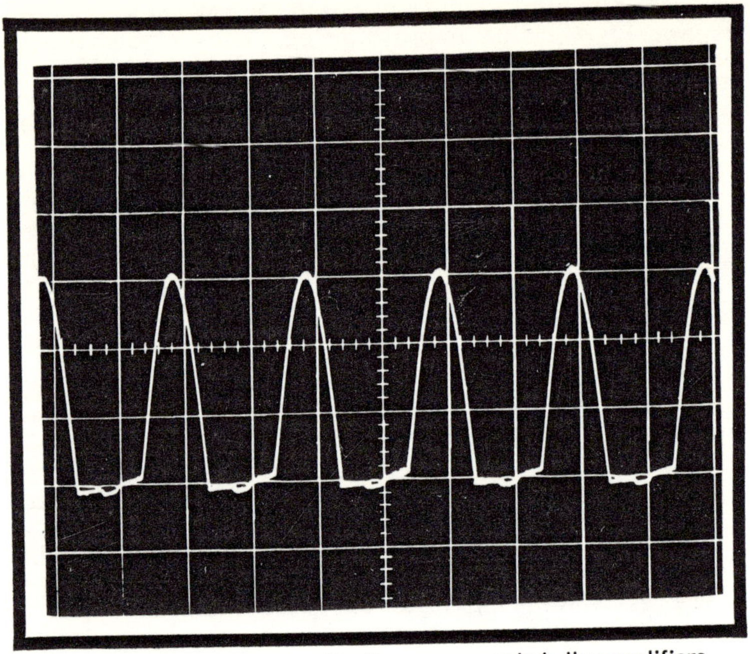

Fig. 14-48. Current delivered by the power supply to the amplifiers.

Thus, when operated at maximum output power, the amplifiers are equivalent to a constant load equal to πR_L connected across each of the two power supplies for split operation, and equal to $2\pi \times R_L$ for single-supply operation.

The situations for the 10W and 60W amplifiers are shown in Fig. 14-49.

HEAT SINKING

When used for audio applications, amplifiers are submitted to erratic speech or music signals. Thus, the average

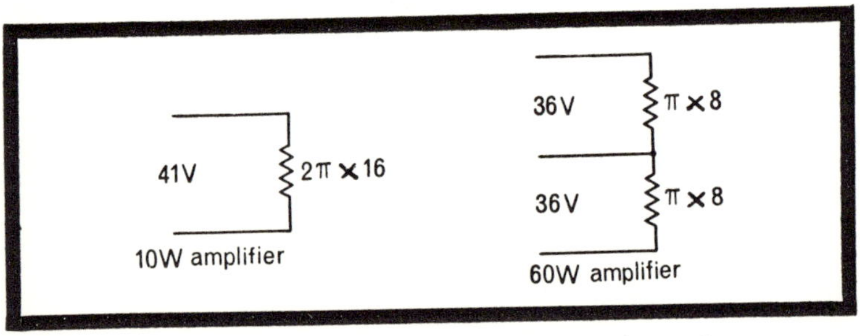

Fig. 14-49. Equivalent loads for the power supplies.

dissipation P_D of the output devices is relatively low, and the limiting characteristic will generally be the second breakdown.

Furthermore, the power supplies normally used are not regulated, so that the continuous output power will be lower than the music power because of a drop of V_{CC}.

A calculation of the heat sinks based upon the maximum dissipation (equal to 14W for each output transistor of the 60W amplifier at $m=2/\pi$ and $\cos\phi=-1$) at the maximum ambient temperature, even under maximum value of V_{CC} (36V for the 60W amplifier), leads to small heat sinks. It is therefore recommended that the heat sink to be designed withstand special conditions (Fig. 14-28), including thermal runaway.

Referring again to Fig. 14-5, consider repetitive pulses of 92W and 3 msec duration, and determine the maximum case temperature at which the limitation would be caused by P_D.

With $\theta_{JC}=1.17$ deg C per watt, $r(t)=0.7$ (normalized transient thermal resistance at a duty cycle of 0.25). It can be found that $T_j=1.17 \times 0.7 \times 92=75$ deg C, and the maximum case temperature $T_{Cmax}=200-75=125$ deg C.

Considering further the use of a heat sink having 3 deg C per watt thermal resistance and a washer of 0.3 deg C per watt, the temperature drop between case and air will be $(3+0.3) \times 92/4=76$ deg C.

Finally, in the above conditions, the maximum ambient temperature $T_{Amax}=200-75-76=49$ deg C.

Chapter 15

Practical Amplifiers From 1 to 100W Output

In the material that follows, a number of amplifiers designed around the previous material are shown in a practical, constructable form. These can be changed to suit individual requirements by using the design details discussed up to this point.

Each of the 1, 3, 10, 20, and 60W amplifiers described in this chapter includes a circuit-board layout with all component values. Table 15-1 gives the parameters for the active devices. The remainder of the amplifiers have circuit diagrams and parts lists giving the parts required for various types of operation.

1W AMPLIFIER

The circuit diagram for the 1W amplifier, Fig. 15-1, shows the inclusion of a 500K potentiometer at the input, which is

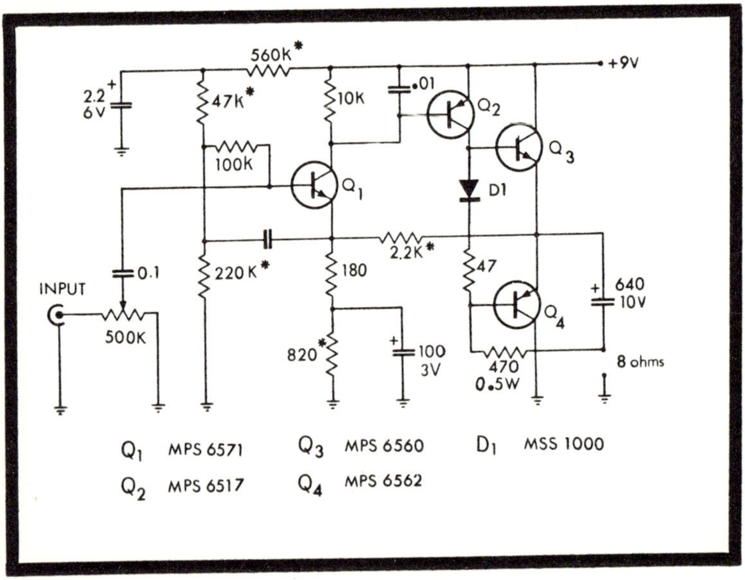

Fig. 15-1. Schematic for 1W amplifier for an 8 ohm load.

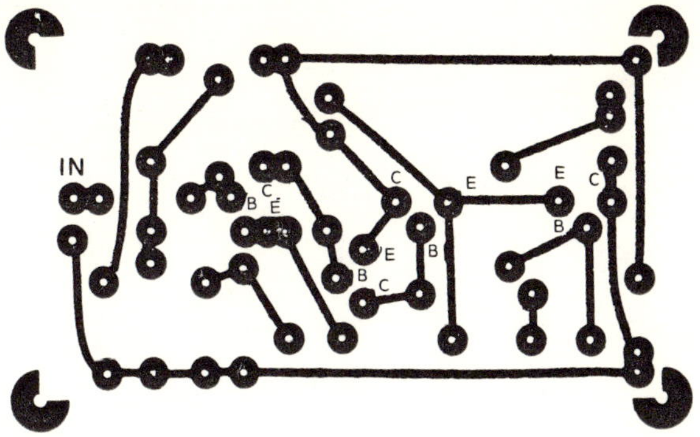

Fig. 15-2. Actual-size 1:1 circuit-board artwork for 1W amplifier.

actually mounted separately from the amplifier. Figure 15-2 is the circuit-board layout. Have this photographed 1 to 1 for use as a negative in making the circuit board.

Typical performance is listed below:

Power: 1W with an 8 ohm load
Harmonic distortion: 10 percent at 1W
Sensitivity: 230 mV input for 1W output
Input resistance: 300K
Frequency response: 3 dB from 40 Hz to 40 kHz
Idling current: 10 mA
Direct current at 1W: 160 mA
Dynamic range: 70 dB (300K source)

1-2W AMPLIFIERS

Figure 15-3 shows a 3-transistor amplifier. The parts list below give the values for each amplifier driving several types of load. No circuit board is included, but it should not be difficult to lay out such a board.

Parts List

Part	1W,8Ω	1W,16Ω	1W, 40Ω	2W, 16Ω	2W,8Ω
V_{cc}	12V	14V	20V	18V	15V
R1	560K	680K	820K	560K	680K
R2	560K	560K	750K	470K	560K
R3	2.2M	2.7M	5.6M	3.3M	2.7M
R5	27	27	33	22	12
R6	560	820	1.0K	220	200
C2 (uF)	250	150	100	150	250
Q2	MPS6560	MPS6561	MPSA05	MPS6560	*MPSU01
Q3	MPS6562	MPS6563	MPSA55	MPS6562	*MPSU51

* Uniwatt.

Performance					
Pout/Load	1W,8	1W,16	1W,40	2W,16	2W,8
Sensitivity at rated	0.8	1.15	0.82	1.2	1.4
Input current at rated	160	110	70	180	240
Idle current, mA	9.0	8.0	9.0	10	17
Input Z, kilohms	560	680	820	560	680
high-frequency cutoff, kHz	20	13	10	14	13

Figure 15-4 is the circuit diagram for a group of 1 and 2W, 4-transistor amplifiers for 8, 16, and 40 ohm loads.

Amplifier	Sensitivity with C4 and R10, mV	Sensitivity without C4 and R10, mV
1W,8	200 mV	600
1W,16	185	800
1W,40	260	1350
2W,16	230	900
2W,8	200	1100

Parts List

Part No.	1W,8Ω	1W,16Ω	1W,40Ω	2W,8Ω	2W,16Ω
Vcc	12V	14V	20V	14V	18V
R3	3.3M	5.6M	6.8M	7.5M	5.6M
R4	1.0K	1.0K	1.2K	1.0K	1.2K
R5	3.3K	3.9K	3.9K	3.9K	3.9K
R6	560	560	1.0K	820	390
R7	27	27	27	18	18
R10	Adjust for desired sensitivity; 200 minimum				
C3(uF)	250	150	100	150	250
Q3	MPS6560	MPS6561	MPSA05	MPSU01	MPS6560
Q4	MPS6562	MPS6563	MPSA55	MPSU51	MPS6562

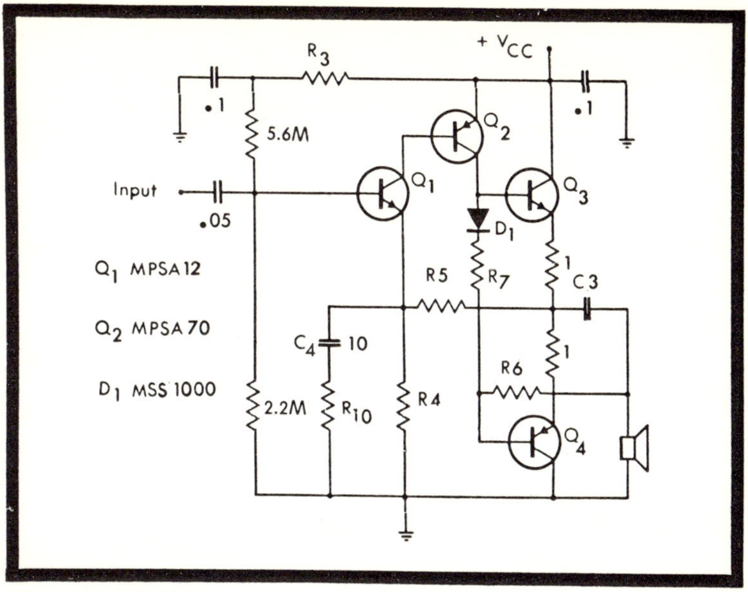

Fig. 15-3. Schematic for 3-transistor, 1 and 2W amplifiers for 8, 16, and 40 ohm loads. (See text for parts values.)

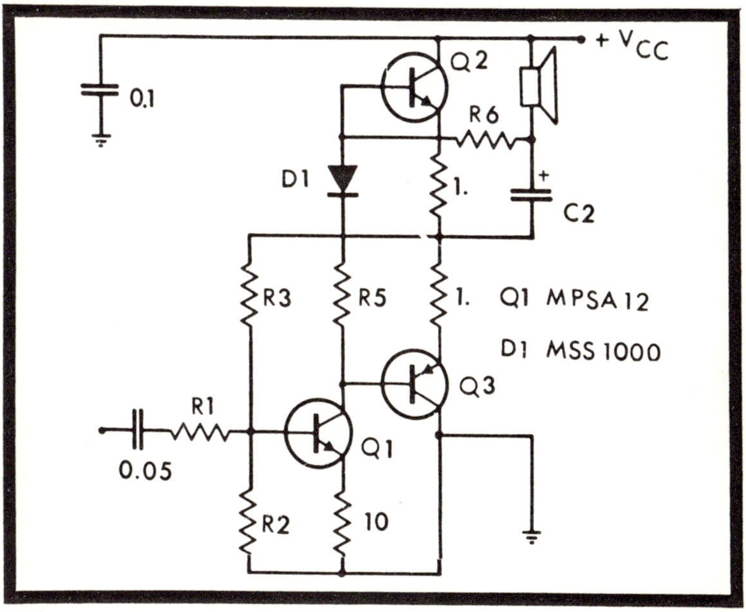

Fig. 15-4. The 1 and 2W, 4-transistor amplifiers. (See text for performance details and parts values.)

PERFORMANCE

P_{out}	1W			2W	
Load (ohms)	8	16	40	8	16
Sensitivity at rated P_{out}, V_{rms}	0.6	0.8	1.35	0.9	1.1
Input current at rated P_{out}, 10 percent THD	180	110	82	240	145
Quiescent idle current, mA	12	13	13	16	15

Total harmonic distortion: 10 percent at rated P_{out}
2 percent at ½ rated P_{out}, measured with requlated power supply

3W AMPLIFIERS

Next is a 3W amplifier driving an 8 ohm load. For this amplifier, circuit-board layout information is included. The performance of this amplifier is a little better, in terms of distortion figures, than that of the amplifiers discussed so far.

Figure 15-5 is the circuit diagram for this amplifier, and Fig. 15-6 is the 1:1 circuit-board layout. Selected (typical) performance characteristics are as follows:

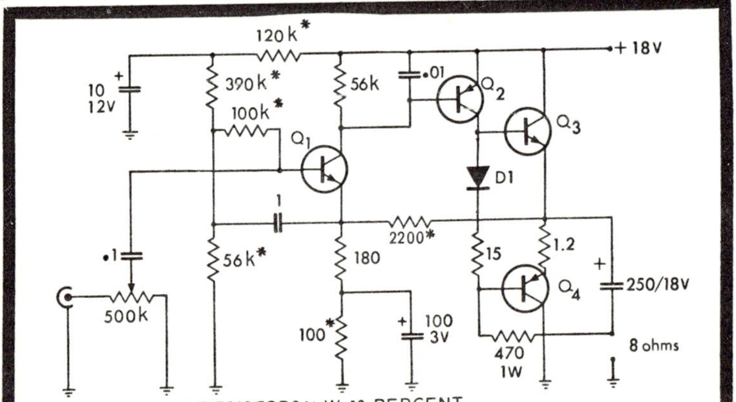

ALL RESISTORS ½ W, 10 PERCENT
* RESISTORS 5 PERCENT CAPACITANCE IN MICROFARADS

Q1	MPS6571	Q3	MPSU 01	D1	MSS 1000
Q2	MPS6517	Q4	MPSU 51		

Fig. 15-5. Circuit for 3W amplifier for an 8 ohm load.

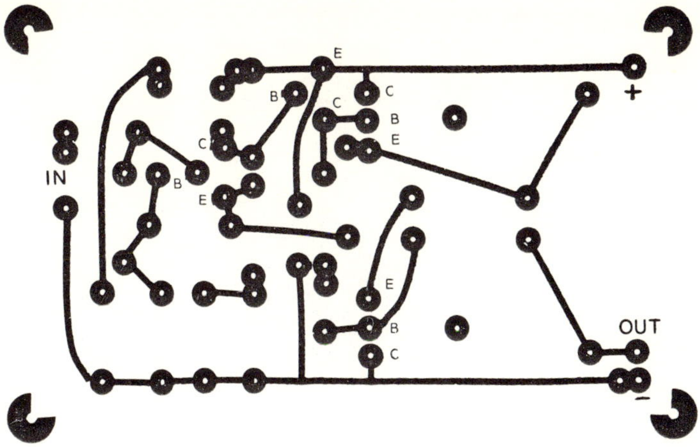

Fig. 15-6. The 1:1 circuit-board artwork for 4-transistor, 3W amplifier.

Power: 3W with 8 ohm impedance
Harmonic distortion: 0.8 percent at 3W
Sensitivity: 380 mV input for 3W output
Input resistance: 300K
Frequency response: 3 dB from 75 Hz to 40 kHz
Idling current: 25 mA
Direct-current at 3W: 290 mA
Dynamic range: 71 dB (300K source)

3 AND 5W AMPLIFIERS

Figure 15-7 is the circuit diagram and parts list for a 3W and 5W amplifier driving an 8 ohm load.

PERFORMANCE	Amplifier Output	
	3W	5W
Idle current, mA	25	25
Current drain at rated power out, mA	290	370
Nominal sensitivity for rated output, mV	250	250
THD at rated maximum power output 50 Hz to 15 kHz, nominal, percent	2	2
THD at rated power output, percent	0.25	0.25
Input impedance, typical, kilohms	300	300
Maximum power output at 5 percent THD		
Without current-limiting diodes, W	4.5	6.7
Without current-limiting diodes, W	4.0	6.0

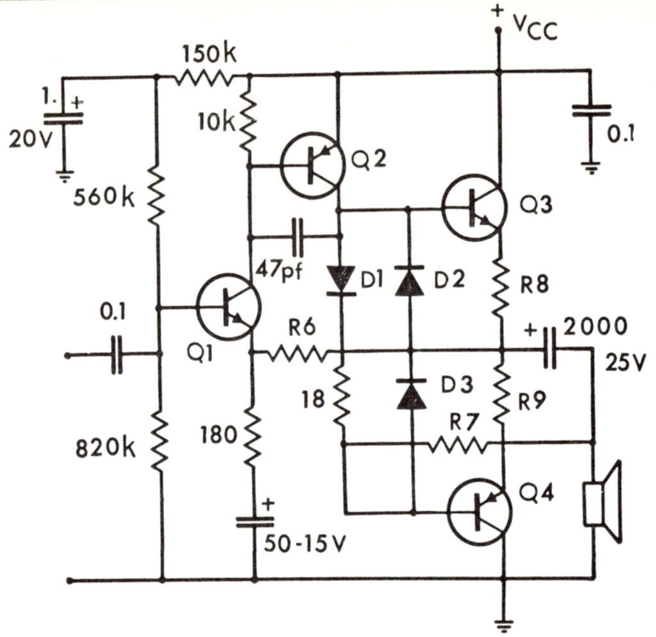

Q1 MPS 6571 Q3 MPSU 01 D1,D2,D3 MSS 1000

Q2 MPSA 55 Q4 MPSU 51

PARTS LIST

	3W	5W
V_{cc}	18V	22V
R6	3.9k	4.7k
R7	470	560
R8	0.75	0.47
R9	0.75	0.47

Fig. 15-7. The 3 and 5W amplifier for an 8 ohm load. Includes current-limiting diodes.

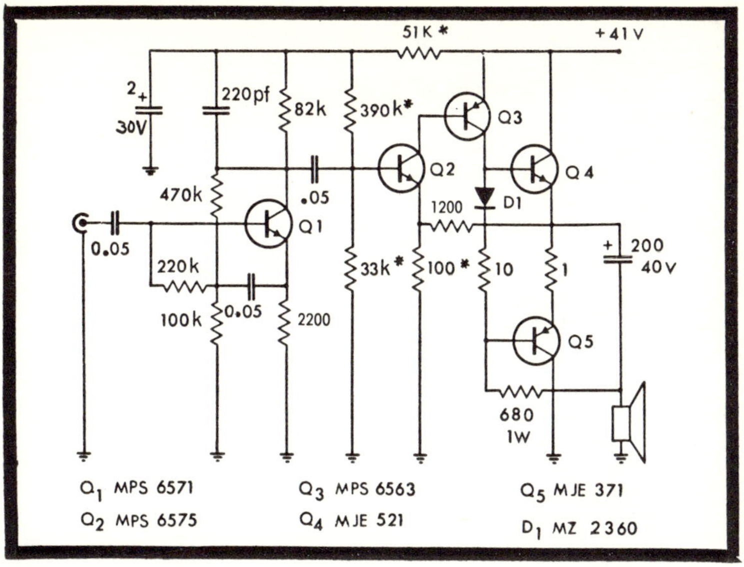

Fig. 15-8. The 10W amplifier for a 16 ohm load.

LOW AMPLIFIER

The 10W amplifier shown in Fig. 15-8 has been mentioned earlier, where some of the design procedures were outlined. Figure 15-9 graphs power output. The circuit-board layout is shown in Fig. 15-10. Selected performance specifications are as follows:

Power: 10W with 16 ohm impedance
Harmonic distortion: 1 percent at 10W
Sensitivity: 0.1V input for 10W output
Input resistance: 250K
Frequency response: 3 dB from 40 Hz to 40 kHz
Idling current, mA: 45
Direct-current at 10W, mA: 390
Dynamic range: 72.5 dB (300K source)

20W AMPLIFIER

Figure 15-11 is the circuit diagram for a 20W amplifier using six transistors. Performance characteristics regarding distortion are materially improved with this design. Figure 15-12 is the circuit-board layout, and the typical performance characteristics are as follows:

Power: 20W with 8 ohm impedance
Harmonic distortion: 0.2 percent at 20W
Intermodulation
distortion: 0.2 percent at 20W
Sensitivity: 0.3V input for 20W output
Input resistance: 100K
Frequency response: 3 dB from 40 Hz to 100 kHz
Idling current, mA 25
Direct-current at 20W, mA: 700

7-35W AMPLIFIERS

In Fig. 15-13, the circuit diagram for a group of 6-transistor amplifiers from 7 to 35W is shown. The blank values should be selected from the parts list that follows. No effort has been made to show great detail or performance because seven configurations are listed.

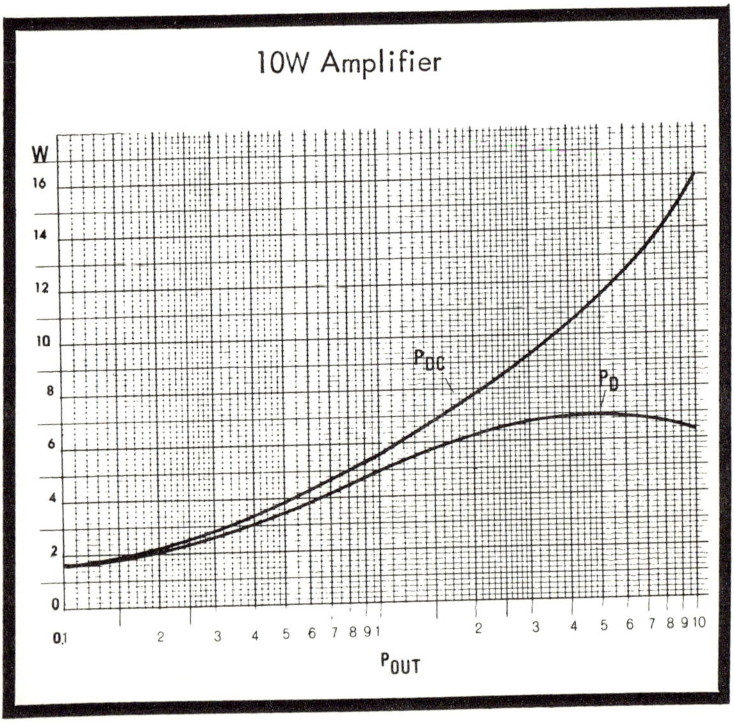

Fig. 15-9. Total dc power P_{dc} and dissipated P_d vs output power P_{out}.

Parts List

Watts	7	10	15	20	25	30	35
Vcc	26V	30V	35V	43V	46V	50V	54V
R1	2.7M	2.7M	560K	560K	220K	220K	220K
R2	1.2M	1.0M	330K	330K	150K	150K	150K
R3	390K	820K	120K	120K	47K	47K	47K
R5	100	82	100	75	220	270	270
R6	8.2K	8.2K	10K	10K	10K	10K	10K
R7	3.9K	3.9K	4.7K	5.6K	5.6K	6.8K	8.2K
R9	0.47	0.47	0.47	0.39	0.39	0.39	0.39
Q2	2N5087	2N5087	2N5087	2N5087	MPSA56	MPSA56	MPSA56
Q3	MPSA05	MPSA05	MPSU05	MPSU05	MPSU05	MPSY05	MPSY05
4	MPSA55	MPSA55	MPSU55	MPSU55	MPSU55	MPSU55	MPSU55
Q5	MJE371	MJE371	MJE105	MJE105	MJE2901	MJE2901	MJE2901
Q6	MJE521	MJE521	MJE205	MJE205	MJE2801	MJE2801	MJE2801
D3-D2	MSS1000	MSS1000	MSS1000	MSS1000	MSS1000	MZ2361	MZ2361
Z_{in}, Kilohms	800	700	200	200	90	90	90
I_{DC}, mA	420	500	580	700	750	850	950
THD, percent	2	2	1		0.5	0.5	0.5
V_{in}, mV	100	100	100	100	450	400	450

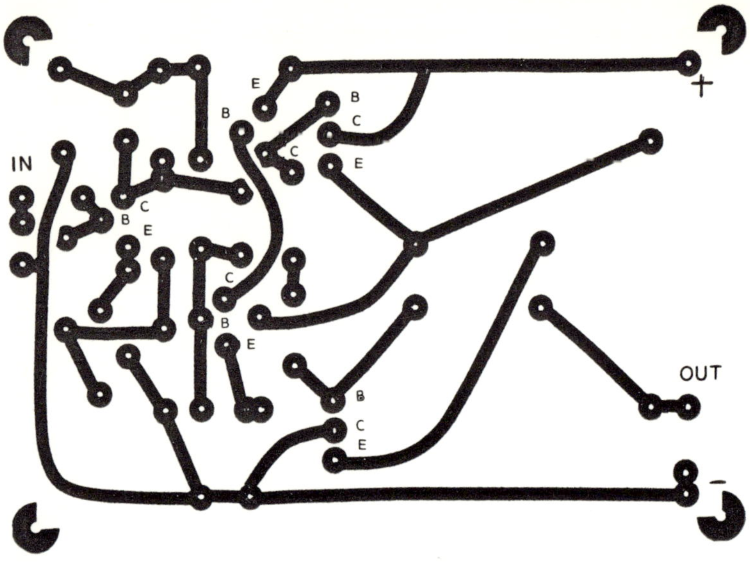

Fig. 15-10. The 1:1 circuit-board artwork for 10W amplifier.

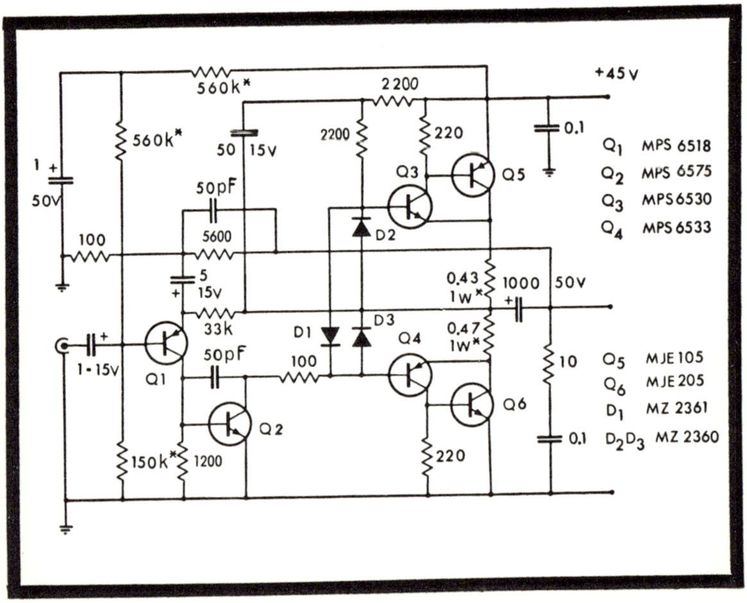

Fig. 15-11. Circuit for 20W amplifier with 8 ohm load.

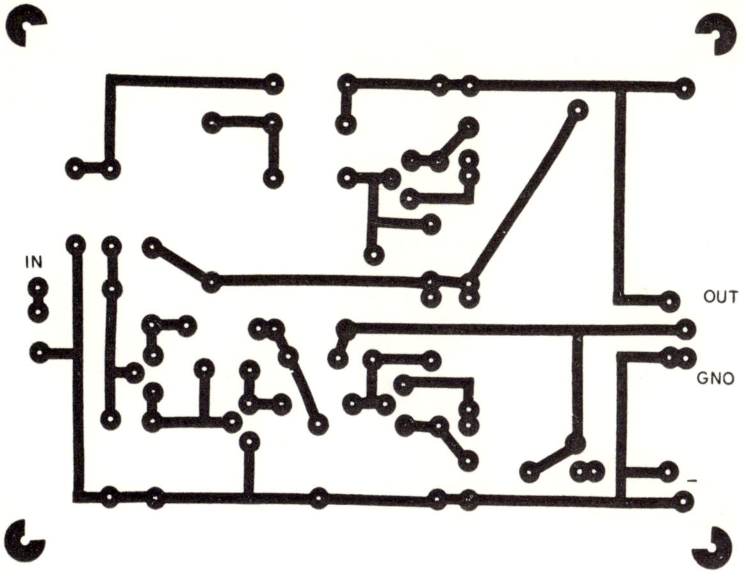

Fig. 15-12. The 1:1 circuit-board artwork for 20W amplifier.

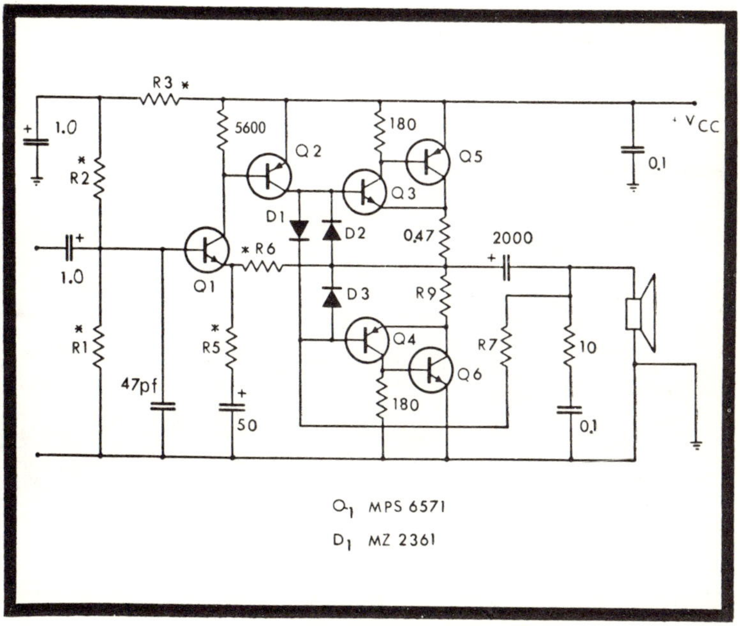

Fig. 15-13. The 7 to 35W amplifiers with 8 ohm load.

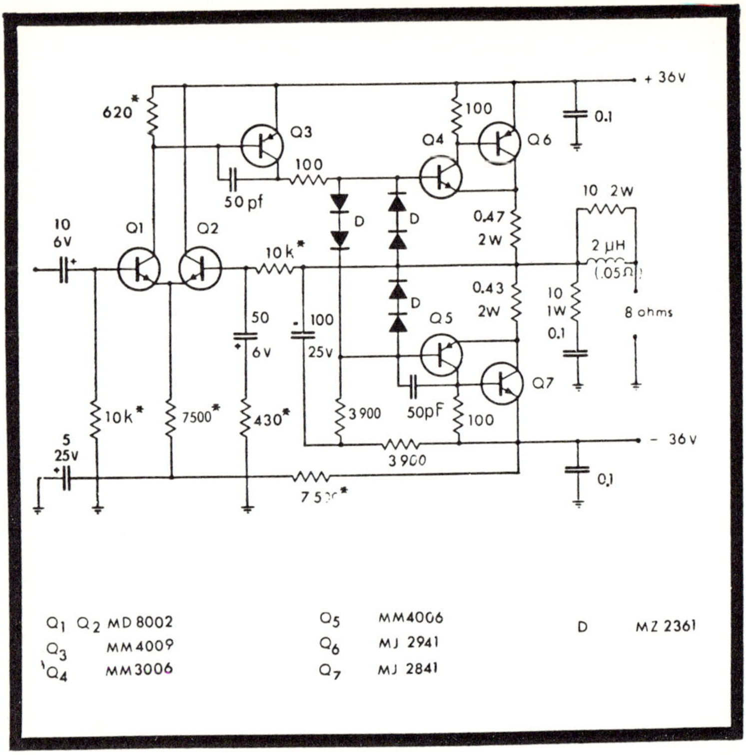

Fig. 15-14. Schematic for 60W amplifier for 8 ohm load, with current-limiting diodes.

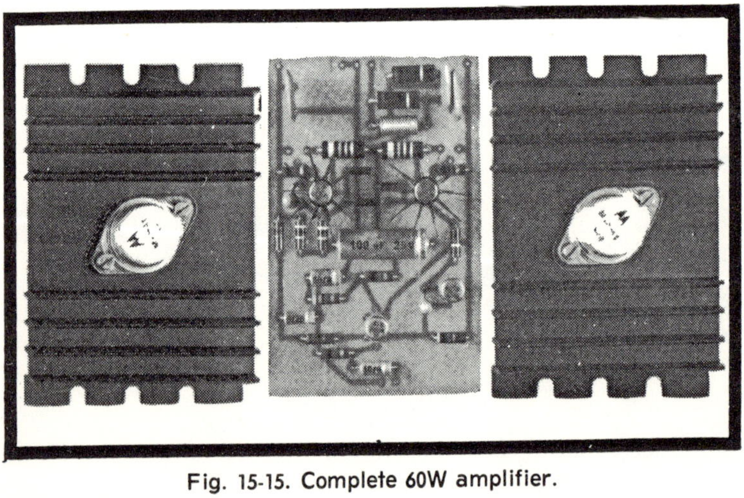

Fig. 15-15. Complete 60W amplifier.

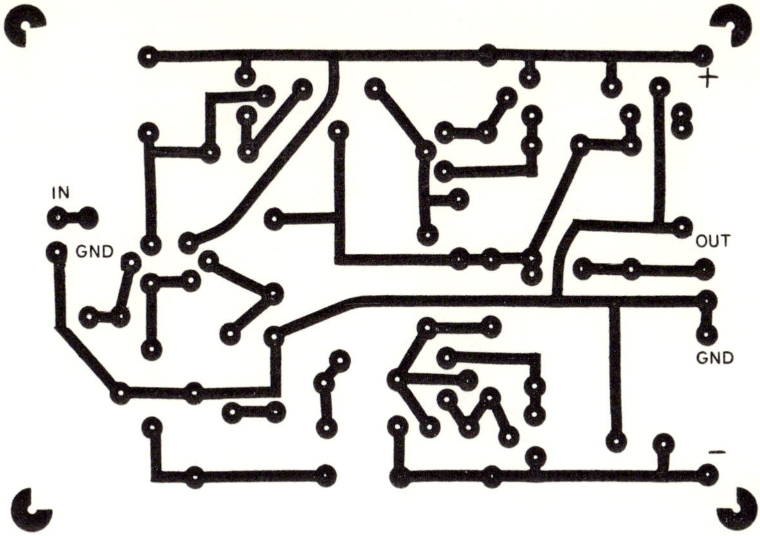

Fig. 15-16. The 1:1 circuit-board artwork for 60W amplifier.

60W AMPLIFIER

The 60W amplifier uses a total of six packages, but the input transistors are a differential pair in the same package; thus, the circuit uses seven transistors to achieve a very good set of performance characteristics. The circuit is current limited with diodes to provide good overload protection.

Figure 15-14 is the circuit diagram. Figure 15-15 gives some details of the final boards as built. Note that two heat sinks are used to ensure adequate cooling of the output transistors. Aslo, note that the driver transistors have a finned heat sink installed for cooling purposes.

Figure 15-16 is the circuit-board layout for the 60W amplifier. Should a stereo amplifier be needed, two circuit boards and four heat sinks will be required. Selected performance characteristics are as follows:

Power: 60W with 8 ohm impedance
Harmonic distortion: 0.1 percent at 60W
Intermodulation
  distortion: 0.1 percent at 60W
Sensitivity: 1V input for 60W output
Input resistance: 10K
Frequency response: 3 dB from 10 Hz to 100 kHz

Idling current, mA: 110
Direct-current at 60W, mA: 1200
Dynamic range: 96 dB (10K source)

35 - 100W AMPLIFIERS WITH POWER LIMITING

Figure 15-17 is the circuit diagram for a series of amplifiers using 11 packages (12 transistors) and incorporating power limiting for short-circuit protection. Details of parts for each type of amplifier are as follows:

RESISTOR VALUES AND POWER SUPPLY VOLTAGES

P_{out} W	Load Ω	R1 ± 5%	R2 ± 10%	R3 ± 5%	R4 ± 5%	R5 ± 5%	R6,R7 ± 5%	R8,R9 ± 10%	R10,R15 ± 5%	R11,R14 ± 5%	R12,R13 ± 5%	V_{cc}
35	4	820	2.7k	18k	1.2k	120	0.39	390	2.7k	1.5k	470	± 21V
	8	560	3.9k	22k	1.2k	180	0.47*	240	3.0k	1.2k	470	± 27V
50	4	680	3.3k	22k	1.2k	100	0.33	360	3.3k	1.5k	470	± 25V
	8	470	4.7k	27k	1.2k	150	0.43*	270	3.9k	1.2k	470	± 32V
60	4	620	3.9k	22k	1.2k	120	0.33	430	3.9k	1.5k	470	± 27V
	8	430	5.6k	33k	1.2k	120	0.39	300	4.7k	1.2k	470	± 36V
75	4	560	4.7k	27k	1.2k	91	0.33	620	5.6k	1.8k	470	± 30V
	8	390	6.8k	33k	1.2k	150	0.39	390	6.8k	1.5k	470	± 40V
100	4	470	5.6k	33k	1.2k	68	0.39	1.0k	8.2k	2.2k	470	± 34V
	8	330	8.2k	39k	1.2k	100	0.39	510	9.1k	1.8k	470	± 45V

Note : All of the above resistor values are in ohms and are 1/2W except for R6 and R7.

All of the amplifiers listed will perform typically as follows:
Output power: Each amplifier will deliver its full rated output power into the nominal load impedance, provided that the power supply has adequate regulation.
Input sensitivity: 1V rms into 10K for full rated output.
Frequency response: Less than 3 dB rolloff from 10 Hz to 100 kHz referenced to 1 kHz.
Power bandwidth: Full rated output power plus or minus 0.5 dB from 20 Hz to 20 kHz.
Total harmonic distortion: Less than 0.2 percent at any power level between 100 mW and full rated output, and at any frequency between 20 Hz and 20 kHz.
Intermodulation distortion: Less than 0.2 percent at any power level from 100 mW to full rated output (60 Hz and 7 kHz mixed 4 to 1).
Damping factor: Over 150 at any frequency from 20 Hz to 20 kHz.

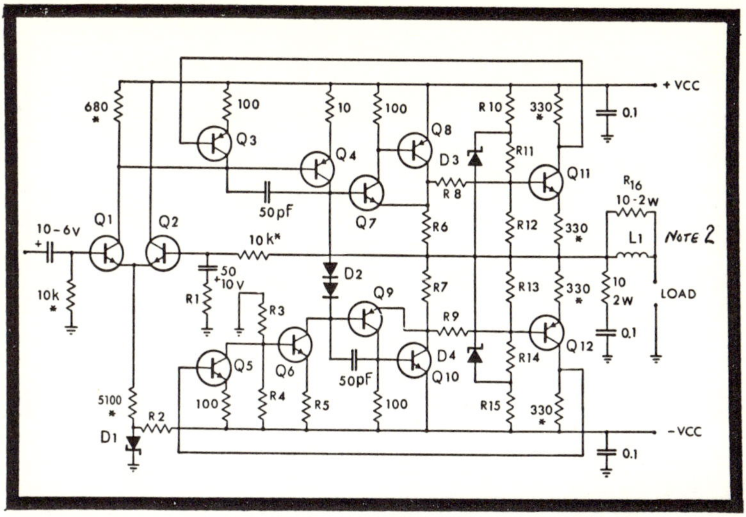

Fig. 15-17. Circuit schematic for 35 to 100W for 4 to 8 ohm loads, with power limiting. (See text for parts values and performance.) **Note** 1: All of the resistors with the values shown are plus and minus 10 percent tolerance, except where ✻ indicates plus and minus 5 percent. 2: L1 is 20 wire close-wound for the full length of resistor R16.

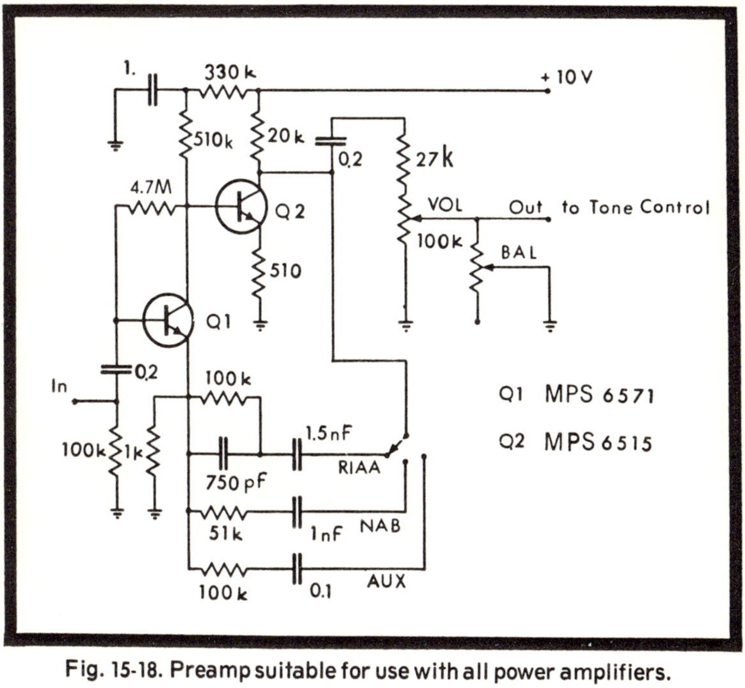

Fig. 15-18. Preamp suitable for use with all power amplifiers.

PREAMP AND TONE CONTROL

Figure 15-18 is the circuit diagram of a preamp suitable for driving any of the amplifiers detailed to this point. The response curves for various modes of operation are shown in Fig. 15-19.

A tone-control circuit can be incorporated as shown in Fig. 15-20, which includes the response curves. Both of these circuits could be incorporated on the same board as the amplifier, if desired. Use the artwork for the amplifier and do the tone control and preamplifier circuitry at 1:1.

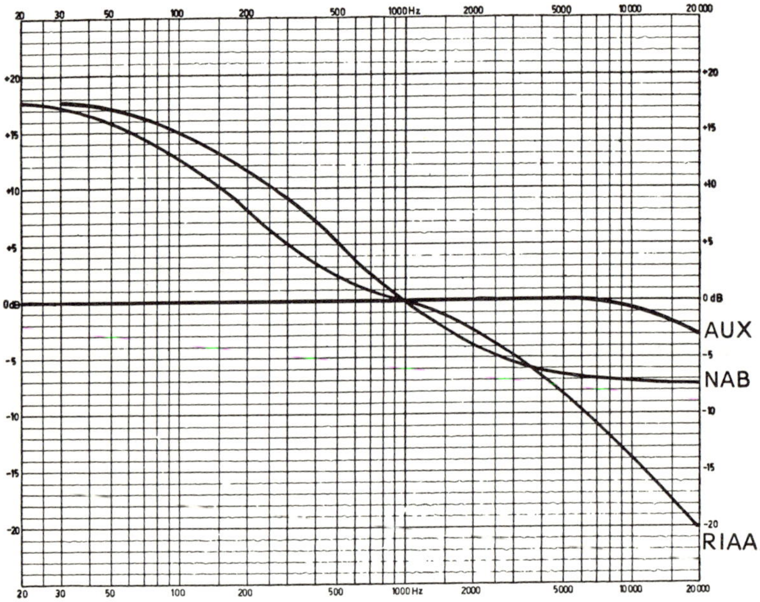

Fig. 15-19. Response curves for preamp.

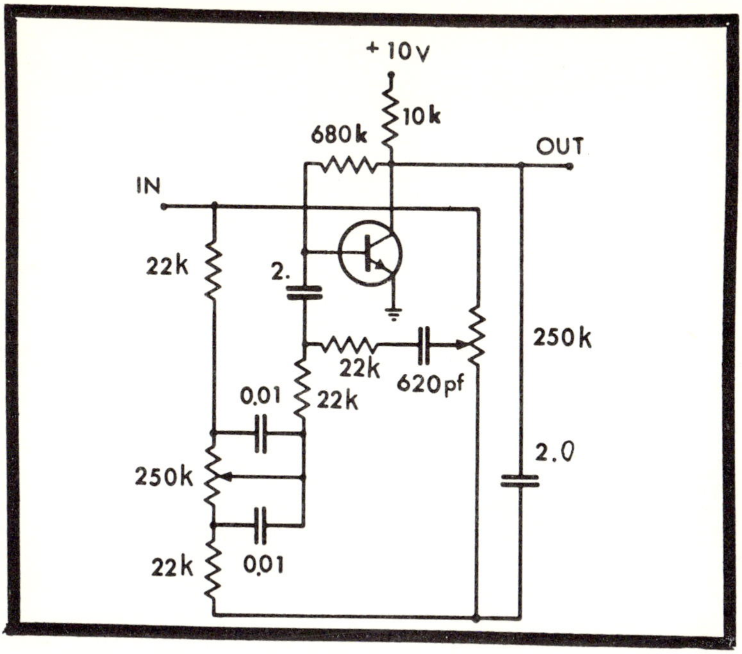

Circuit performance:
Range of bass control: plus or minus 15 dB at 50 Hz. Range of treble control: plus or minus 11 dB at 15 kHz. THD: 0.2 percent from 50 Hz to 15 kHz at 0.5V rms.

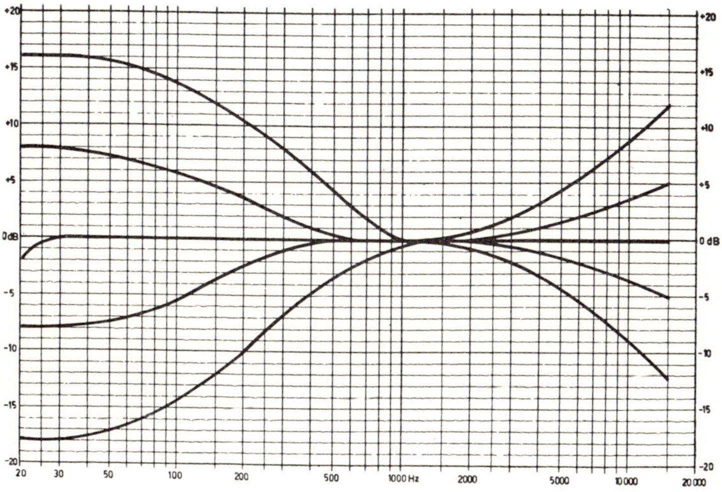

Fig. 15-20. Tone control schematic and response curve. (Capacitance in microfarads.)

Table 15-1
Output Complementary Pairs for Audio-Amplifiers

P_o	4 Ω LOAD				8 Ω LOAD				16 Ω LOAD			
W	V_{pk} V	I_{pk} A	V_{cc} Ap. V	Recomm. pair	V_{pk} V	I_{pk} A	V_{cc} Ap. V	Recomm. pair	V_{pk} V	I_{pk} A	V_{cc} Ap. V	Recomm. pair
1	2.8	0.7	7.5	MPS-6560 MPS-6562	4	0.5	9	MPS-6560 MPS-6562	5.6	0.35	14	MPS-6561 MPS-6563
2	4.0	1.0	9	MPS-U01 MPS-U51	5.7	0.7	14	MPS-6560 MPS-6562	8	0.5	18	MPS-6560 MPS-6562
3	4.9	1.24	12	MPS-U01 MPS-U51	6.9	0.87	18	MPS-U01 MPS-U51	9.4	0.6	22	MPS-U01 MPS-U51
5	6.3	1.6	18	MJE-105 MPS-U01 MJE-205 MPS-U51	8.9	1.12	22	MPS-U01 MPS-U51	12.6	0.79	30	MPS-U01 MPS-U51

10	8.9	2.24	22	MJE-105 MJE-205	12.6	1.6	30	MJE-105 MJE-371 MJE-205 MJE-521	17.8	1.12	41	MJE-371 MJE-521
20	12.6	3.16	30	MJE-2801 MJE-2901	17.9	2.23	45	MJE-105 MJE-205	25.2	1.6	60	MJE-2801 MJE-2901
35	16.7	4.2	40	MJ-2841 MJ-2941	23.7	2.95	54	MJE-2801 MJE-2901	33.4	2.1	75	MJ-2841 MJ-2941
60	21.9	5.5	50	MJ-802 MJ-2841 MJ-4502 MJ-2941	31	3.87	72	MJ-2841 MJ-2941	43.8	2.75	100	MJ-802 MJ-4502
100	28.3	7.1	65	MJ-802 MJ-4502	40	5	90	MJ-802 MJ-4502	56.6	3.55	120	2N5633

P_o = Output power

I_{pk} = Peak output current

V_{pk} = Peak output voltage

V_{cc} = Total supply voltage

Part III

Linear ICs

Chapter 16

Linear Integrated Circuits

Six linear integrated circuits bear discussion in this book because of their wide acceptance in the industry and their consequent availability. These are the 702 wideband amplifier, the 709 operational amplifier (op amp), the 710 and 711 comparators, and the 741 and 748 op amps.

The 702 wideband amplifier is most easily available in either a T099 can or a dual-inline package. It has a requirement for unbalanced positive and negative power supplies, which are normally plus 12V and minus 6V. The 702 can be operated from a single power supply, with consequent restriction on the output swing.

The 702 is described in the literature as a high-gain, wideband op amp with differential inputs and a single-ended, emitter-follower output. The circuit must be externally compensated; the appropriate data sheet will include information on this compensation. For our purpose, a few circuits will be discussed below, and the compensation will simply be indicated on the diagram with no effort to go into the methods of determining proper compensation.

The 709 is described as a high-performance op amp with high-impedance differential inputs and a low-impedance output. This amplifier can be operated from an op amp power supply (plus and minus 15V on either side of ground), and it, too, must be externally compensated. Both this amplifier and the 702 require, in addition to proper compensation, a small (0.01 uF) capacitor near each power input terminal to help eliminate any tendency to oscillate.

The greatest practical problem encountered with these devices is their tendency to oscillate at some arbitrary frequency that has nothing to do with the desired output. Thus, considerable care should be taken to ensure both proper compensation and power-input decoupling.

The 710 and 711 devices are specially designed to operate as comparators interfacing external circuitry to 5V logic. The output is a logical "0" or "1," depending upon the condition of the input circuit.

The 741 was developed to overcome, to some extent, the requirement for external compensation in these amplifiers. It incorporates improved circuitry, is a true op amp, and has

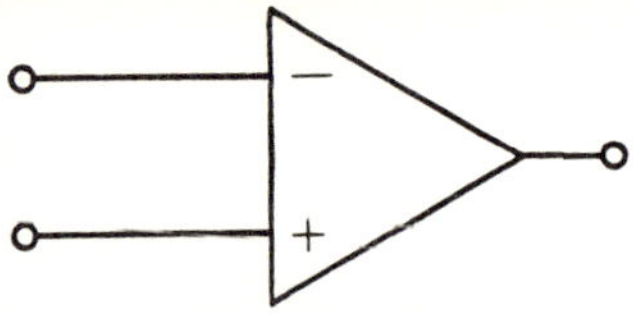

Fig. 16-1. Basic op amp symbol.

internal compensation. The 741 has a slewing-speed limitation, however; a new device, the 748, is a 741 with external compensation required.

THE OP AMP

The op amp is an unusual device. It is also a theoretical designer's "dream" because it works exactly as it is supposed to, with very minor exceptions. The mathematics of the circuit in use are not for this book; however, some practical points are worth discussing.

First, the lowest gain figure (open-loop) quoted for a 702 is 1000, and the highest is 7000. For the 709, these figures are from 10,000 to 45,000! Taking another figure for the 709, the maximum output voltage swing is from plus 14V to minus 14V or a total of 28V with a plus and minus 15V (30V overall) power supply.

The interesting point, however, is that to achieve a 28V swing at a gain of 10,000 requires only a 2.8 mV input! And, at a gain of 45,000, it requires only a 622 uV input! Theoretically, an op amp should have infinite gain and zero output impedance. This would make the input resemble an infinite impedance. But this is impossible to achieve practically with present-day technology, so we must settle for something less.

As mentioned, the mathematics for using this device are quite involved; as a practical start, however, the op amp is used only occasionally in an open-loop configuration. A feedback circuit is normally incorporated.

Figure 16-1 shows the basic symbol for an amplifier and the way the symbol is "signed" for op amps. The plus- and minus-signed input terminals (also called noninverting and inverting inputs, respectively) are shown on the left, and normal convention **usually** places the minus or inverting terminal at the top.

The unity-gain amplifier in Fig. 16-2 is a normal circuit configuration for a 709 op amp. The 470 pF capacitor from pin 6 to pin 5 is the "output compensation" capacitor. The input compensation network (1K and 0.005 uF) is connected from pin 1 to pin 8. Pin 4 is the negative supply connection (not shown), and pin 7 is the positive supply connection (also not shown). Each of these should have a 0.01 uF ceramic capacitor connected from a point as near the pin as possible to ground.

The final pins are the input and output. In this amplifier, pin 3, the noninverting input (plus sign), is connected to ground through a 1K resistor to balance the input resistance from pin 2 to the signal input. Pin 2 is now called the summing input and, theoretically, it should always remain at ground potential.

Input and output signals are shown in Fig. 16-3 for a 2 kHz square wave of about 0.28V amplitude. The ratio of $R_F:R_{in}$ is 1:1, and this results in a gain of 1 or unity; thus, the output is also 0.28V!

In Fig. 16-4, the input frequency has been increased to 70 kHz. At this point, there is a tendency for the output to ring; also, the output rise and fall times have been degraded by slewing-speed restrictions. Figure 16-5 shows the result of removing the output-compensation capacitor.

Figure 16-6 shows the waveforms obtained at 2 kHz with R_F equal to 10K. Gain for the op amp is found from R_F divided by R_{in}. In this case, the gain is 10K divided by 1K or 10. The output is now 2.8V and, again, the rise and fall times are degraded. In Fig. 16-7, R_F has been increased to 100K, making a "gain of 100" amplifier, and the circuit is being overdriven with 0.28V input.

In Fig. 16-8, the input has been reduced to 28 mV. The rise and fall degradation is even more apparent.

The waveforms in Figs. 16-9 and 16-10 match those in Figs. 16-3 and 16-4, respectively, but a 741 op amp has been sub-

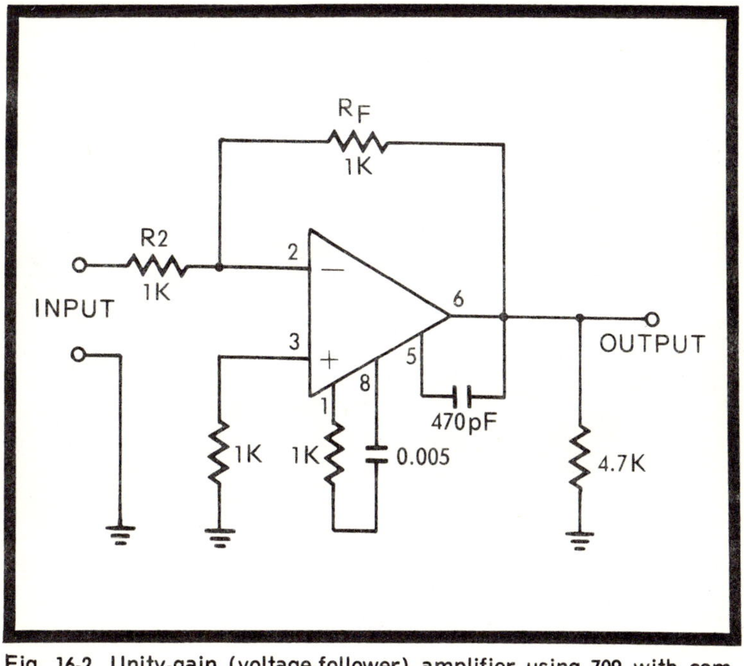

Fig. 16-2. Unity-gain (voltage-follower) amplifier using 709 with compensation.

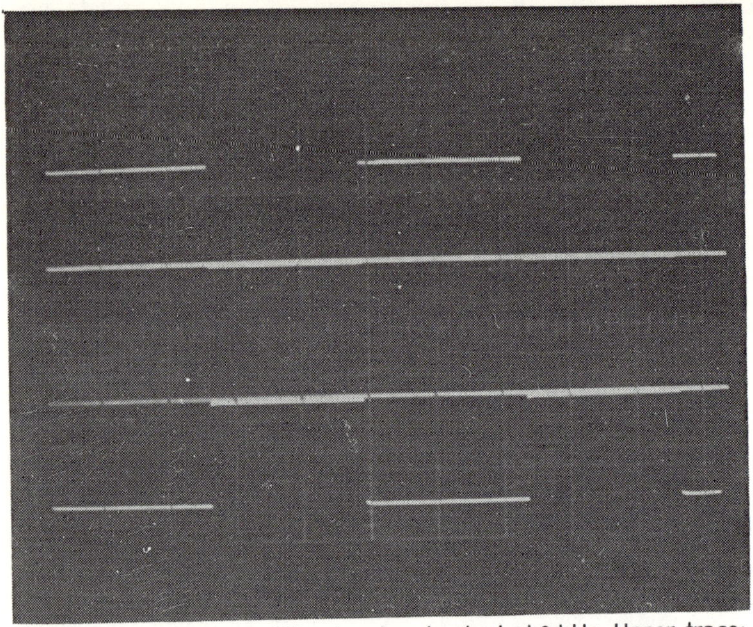

Fig. 16-3. Unity-gain amplifier input and output at 2 kHz. Upper trace: Input 0.2V per cm. Lower trace: Output 0.2V per cm. Time base: 0.1 msec per cm.

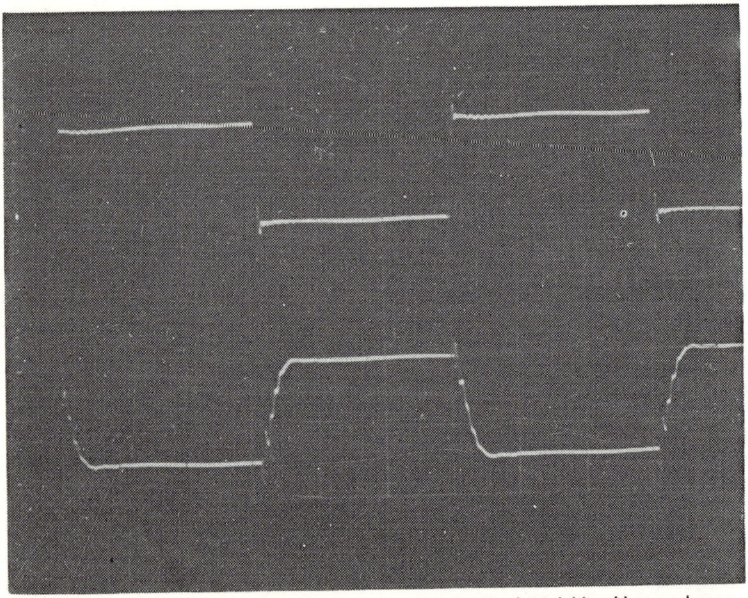

Fig. 16-4. Unity-gain amplifier input and output at 80 kHz. Upper trace: Input 0.2V per cm. Lower trace: Output 0.2V per cm. Time base: 2 msec per cm.

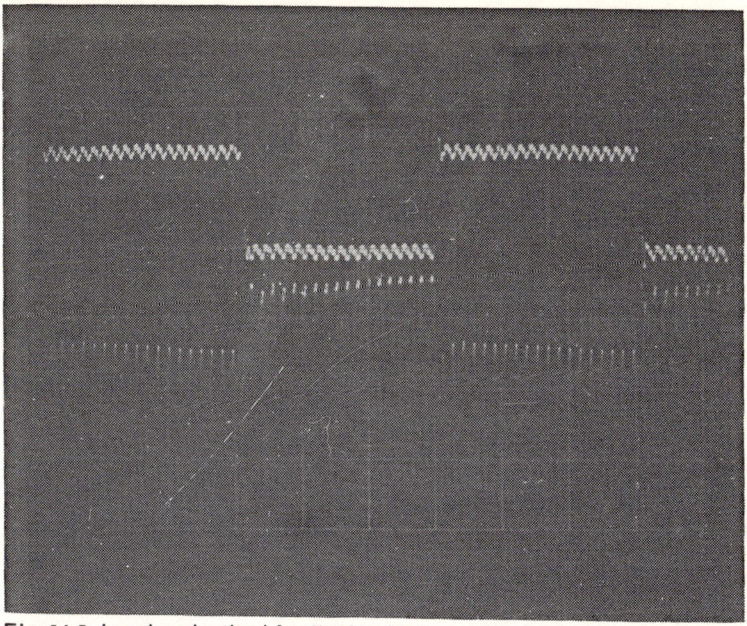

Fig. 16-5. Input and output for "gain of 10" amplifier. Upper trace: Input 0.2V per cm. Lower trace: Output 2.0V per cm. Time base: 0.1 msec per cm.

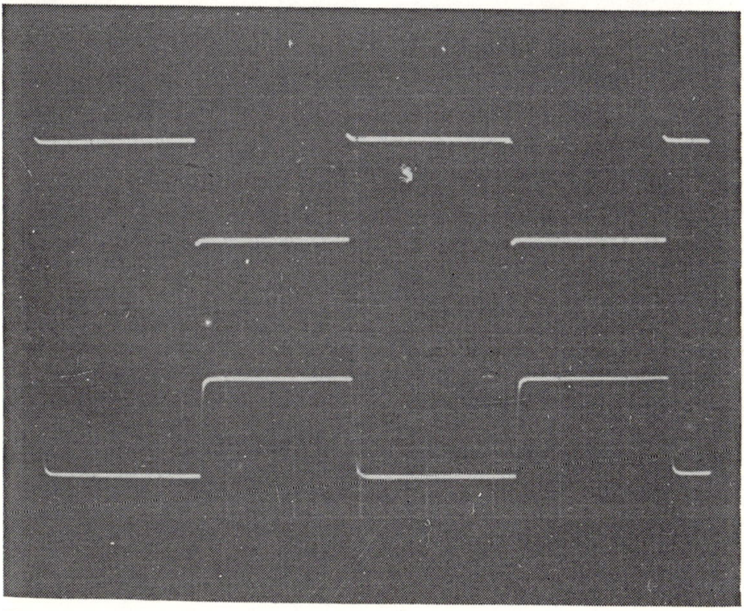

Fig. 16-6. Input and output of unity-gain amplifier with output compensation removed. Ringing is about 3 MHz.

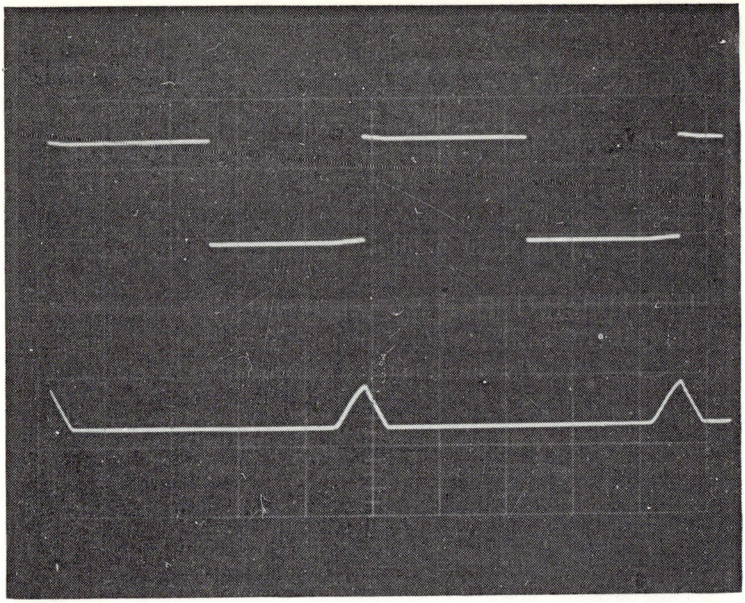

Fig. 16-7. Input and output for "gain of 100" amplifier (overdriven). Upper trace: Input 0.2V per cm. Lower trace: Output 20V per cm. Time base: 0.1 msec per cm.

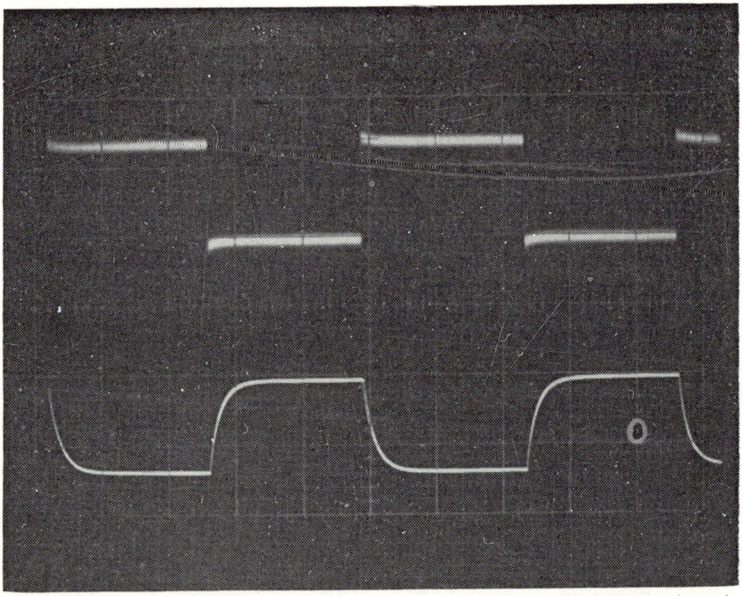

Fig. 16-8. Input and output for "gain of 100" amplifier with input reduced. Upper trace: Input 20 mV per cm. Lower trace: Output 2.0V per cm. Time base: 0.1 msec per cm.

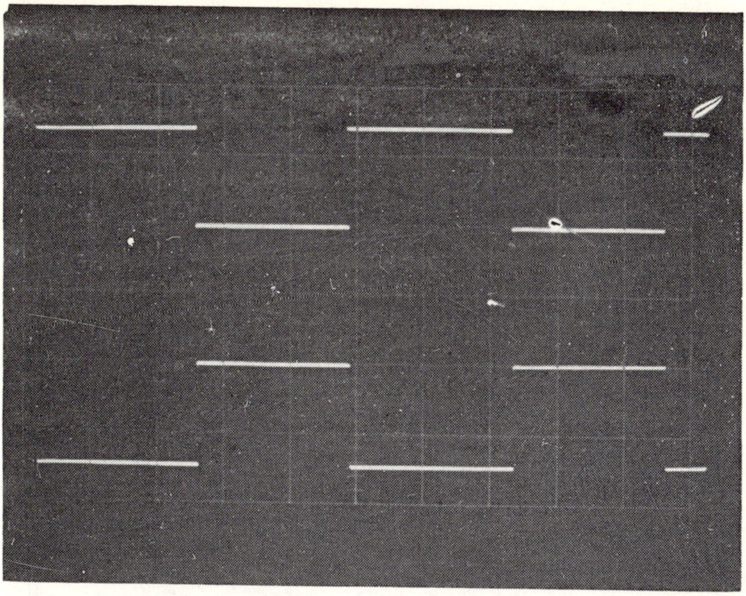

Fig. 16-9. Input and output of unity-gain amplifier using 741 op amp and 2 kHz input. Settings as for Fig. 16-3. Input and output of unity-gain amplifier using 741 op amp and 80 kHz input.

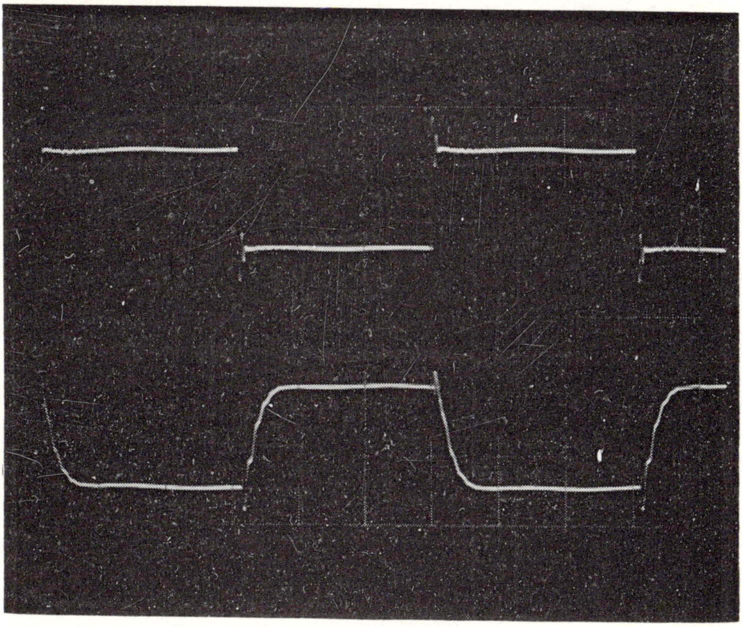

Fig. 16-10. Note similarity to Fig. 16-4.

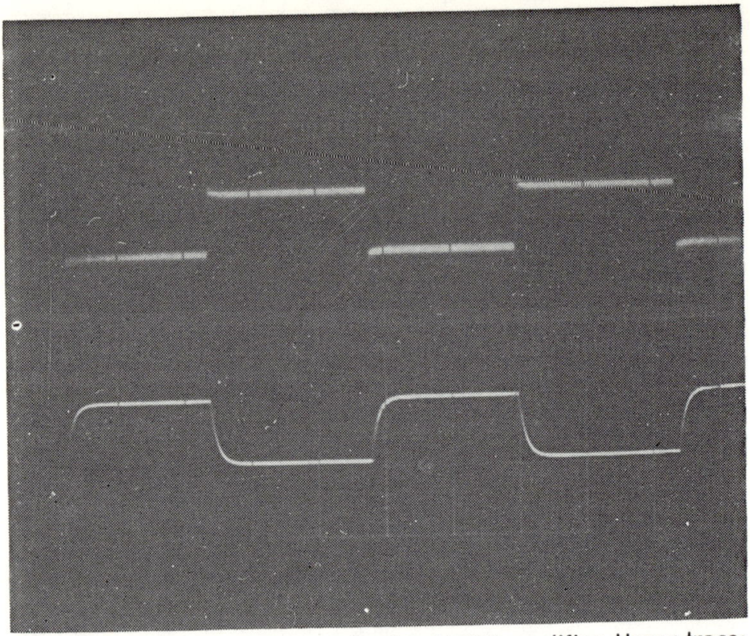

Fig. 16-11. Input and output for a "gain of 1000" amplifier. Upper trace: Input 10 mV per cm. Lower trace: Output 10V per cm. Time base: 2 msec per cm.

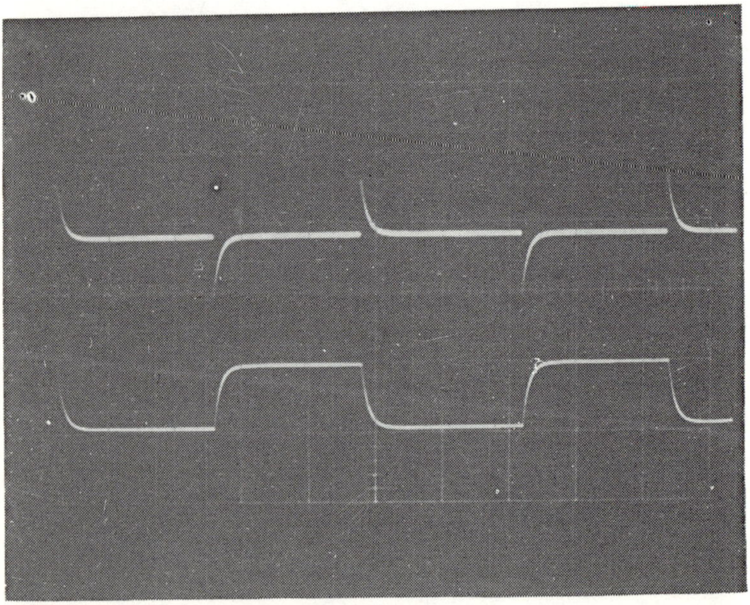

Fig. 16-12. Input and output for a "gain of 1000" amplifier, but with upper trace taken from summing point (gain 2).

stituted. Finally, with R_F equal to 1M, making a "gain of 1000" amplifier, the input frequency must be reduced to 100 Hz to achieve the results shown in Fig. 16-11.

In Fig. 16-12, the upper trace is taken from the summing point to illustrate the reason for the degradation of the output waveform. Finally, Fig. 16-13 shows the circuit diagram for a variable-gain amplifier using a 741 op amp.

The principle is quite simple. If the variable resistor is set at 10K, the amplifier has a gain of 10. At 20K, the gain will be 20. At 100K, the gain will be 100. Therefore, the gain of this particular amplifier can be varied from, say, 0 to 100.

Using a 1M potentiometer, the gain could be varied from 0 to 1000. Again, as the gain is increased, the upper frequency limit is decreased.

Figures 16-14, -15, -16, -17 shows various ways in which the 702 wideband amplifier can be used. The photodiode amplifier in Fig. 16-14, for example, could be used in a computer card reader; it could also be connected for use as a can counter on a production line or in a security alarm system.

The circuits of Figs. 16-15 and 16-17 are similar. Both will operate at radio frequencies, making them useful as videotape-head amplifiers or as i-f or rf amplifiers. In the precision ramp generator of Fig. 16-16, the capacitor will charge in response to the positive voltage applied, generating a ramp in the process. The switch across the capacitor is used to discharge the capacitor in order to start a new cycle. This switch can be mechanical or electronic (transistor or FET), if desired.

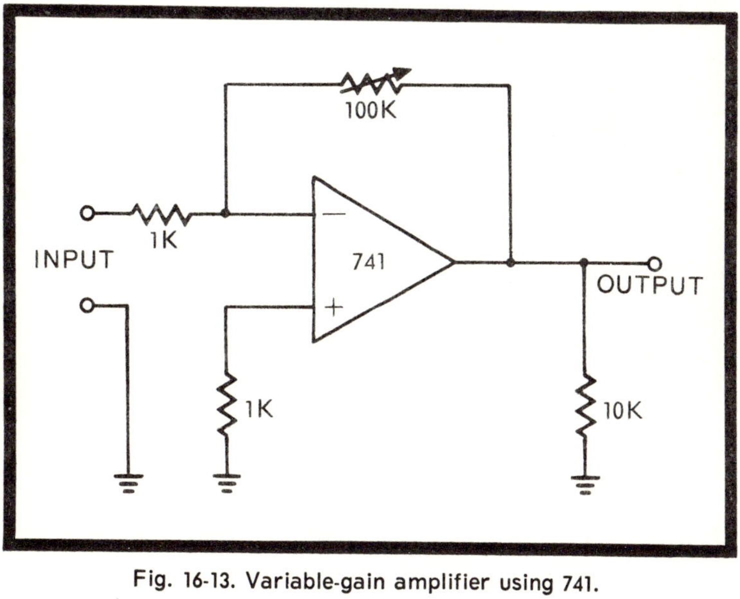

Fig. 16-13. Variable-gain amplifier using 741.

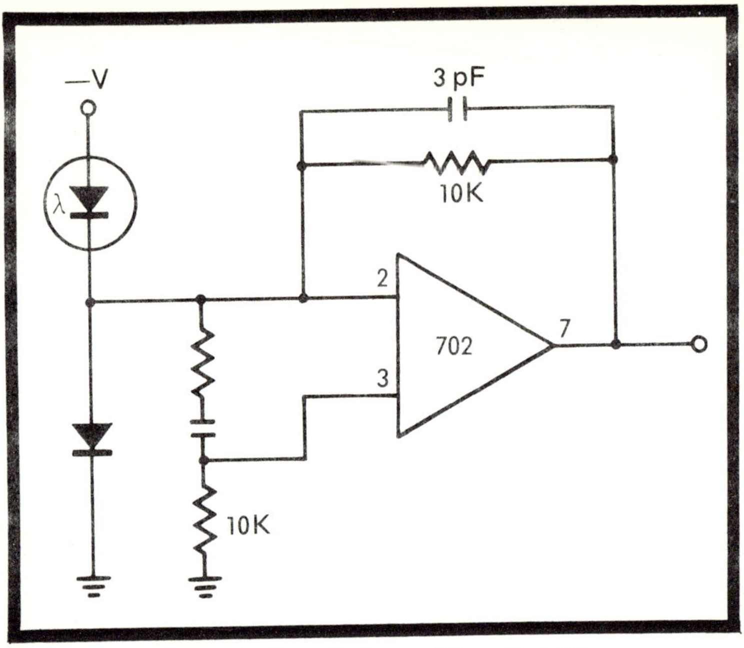

Fig. 16-14. Photodiode amplifier using 702.

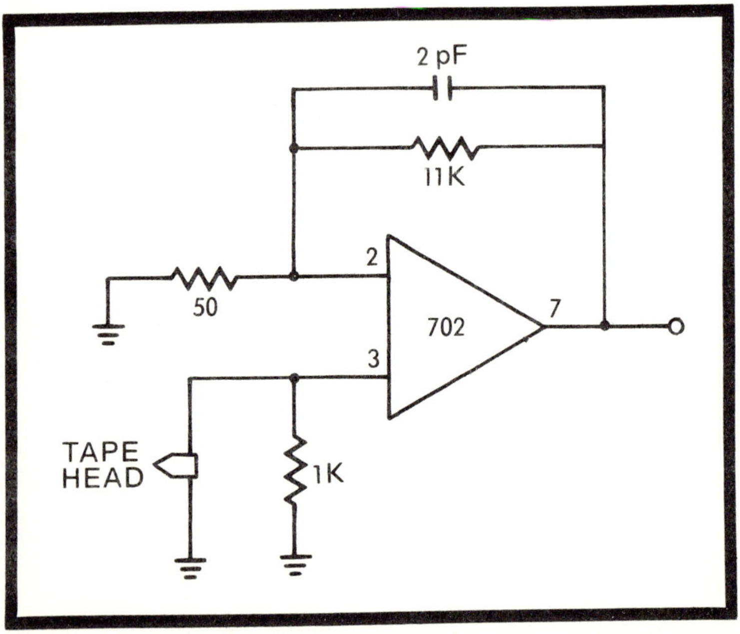

Fig. 16-15. Tape-head amplifier with video response.

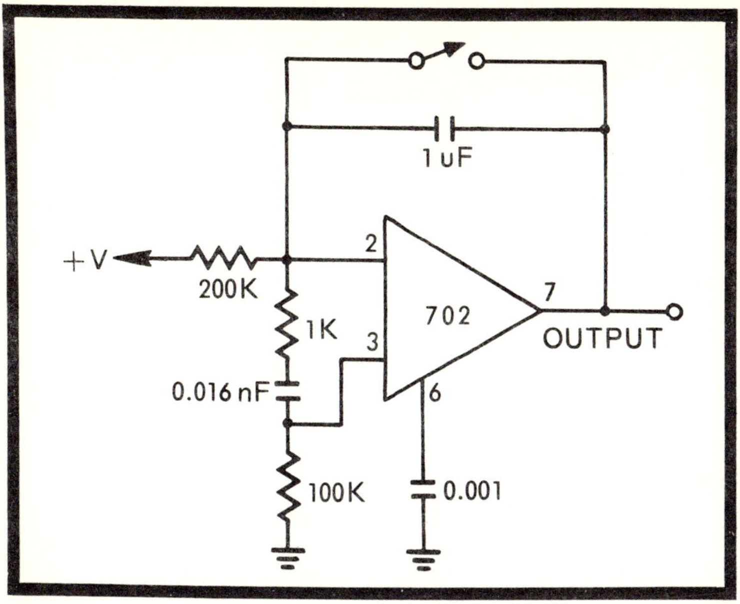

Fig. 16-16. Precision ramp generator using 702.

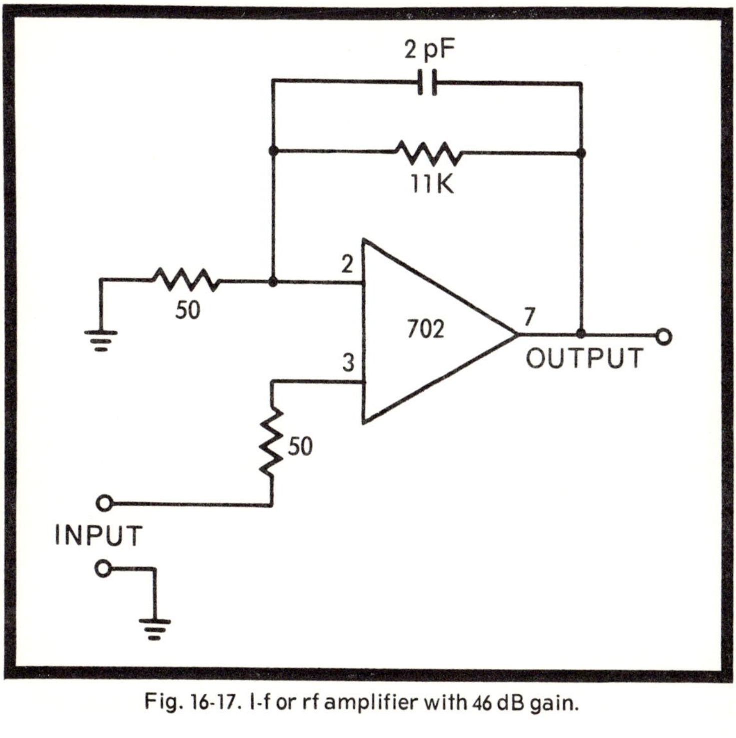

Fig. 16-17. I-f or rf amplifier with 46 dB gain.

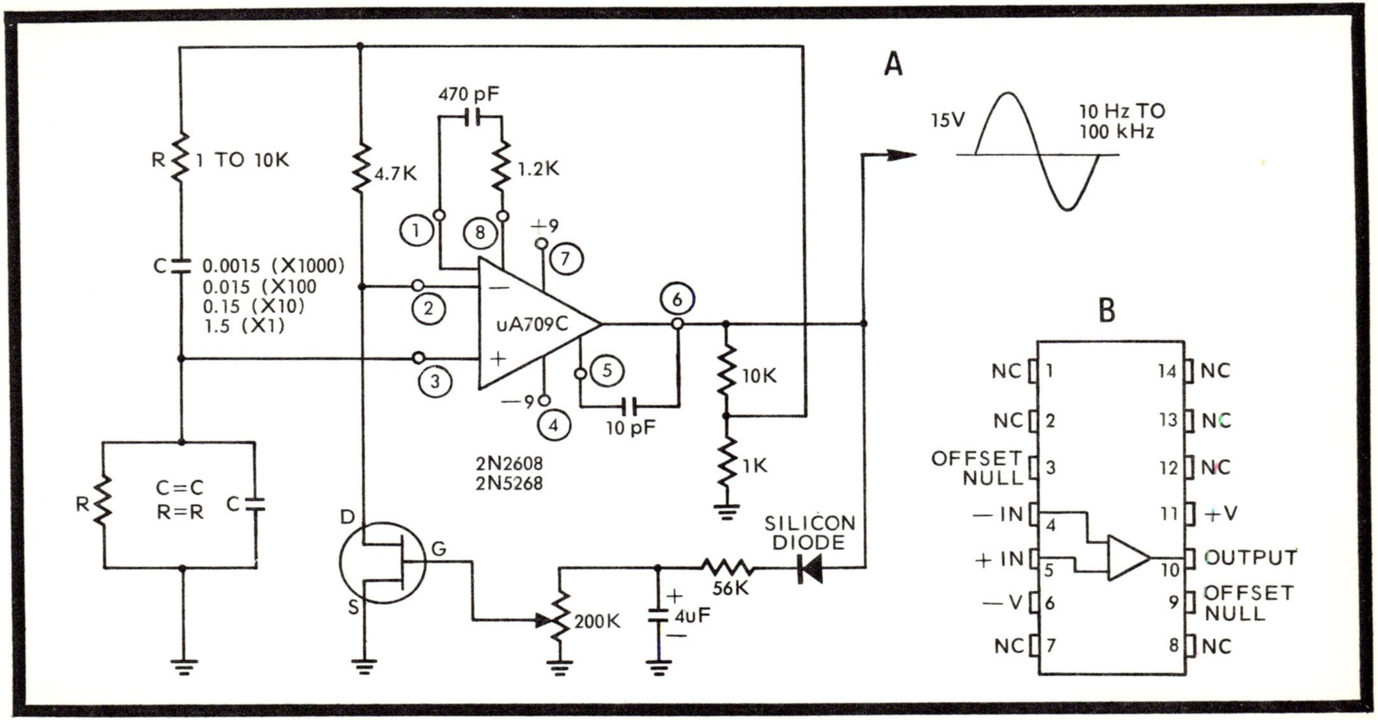

Fig. 16-18. Audio oscillator with FET level control. In A, the schematic; in B, connections to chip.

OSCILLATORS

It is possible to build very effective oscillators with op amps. The circuit of Fig. 16-18A is one possibility. A Wein bridge network is used for frequency control. With a 1-110K range for resistance R and the values of capacitance C shown, the range is from 10 Hz to 100 kHz.

A unique feature of this oscillator is the method used to maintain amplitude control. The normal system uses a thermistor or lamp to alter the feedback and control the output level. This oscillator uses a FET. A portion of the output is rectified and used as a dc bias on the gate of the FET to control FET resistance. An increase in signal amplitude will cause FET resistance to change and reduce the system gain, bringing the output back to the set level. The dual-in-line 709 package is shown in Fig. 16-18B.

INTEGRATORS

Mathematically, integration has to do with summation. The op amp integrator integrates the input, and gives the result of the integration as an output; the practical side, however, is equally useful.

In Fig. 16-19, the integrator shown will convert a 100 Hz square wave to a triangular wave, as shown in Fig. 16-20. Note that each time the input changes polarity, the triangular wave changes direction.

The entire effect comes from the linear charge and discharge of the feedback capacitor. Also, the signal source for this input is a square wave that goes from zero volts in a positive direction only; thus, a large capacitor is required at the input to keep the tops of the square wave flat.

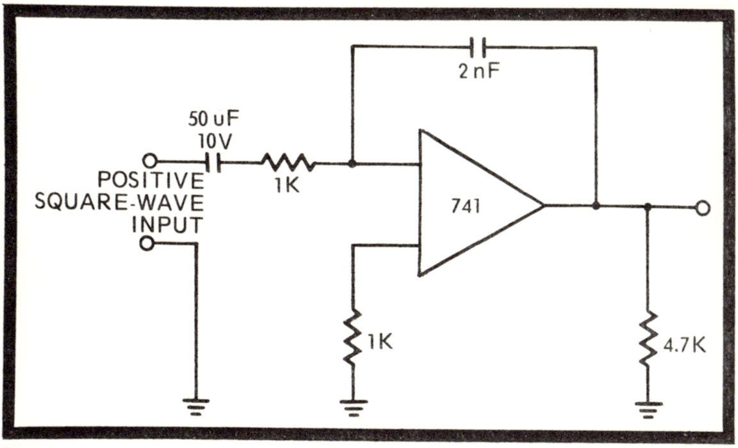

Fig. 16-19. Integrator used as a square wave to triangular wave converter.

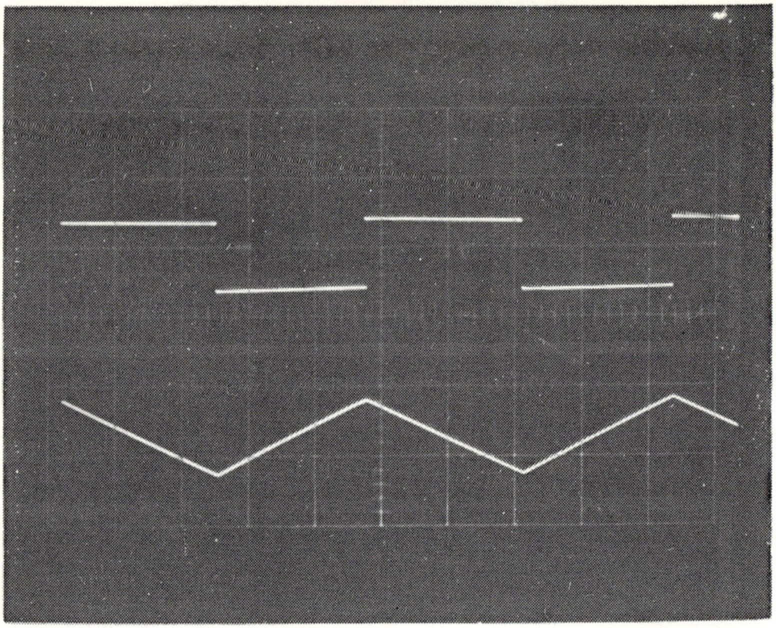

Fig. 16-20. Input and output of square-to-triangular converter—100 Hz. Upper trace: Input 1V per cm. Lower trace: Output 1V per cm. Time base: 2 msec per cm.

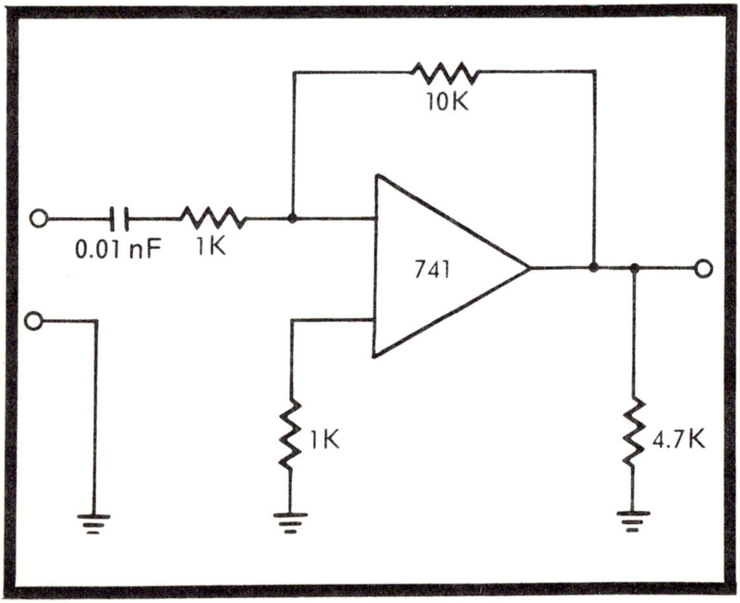

Fig. 16-21. Differentiator using 741.

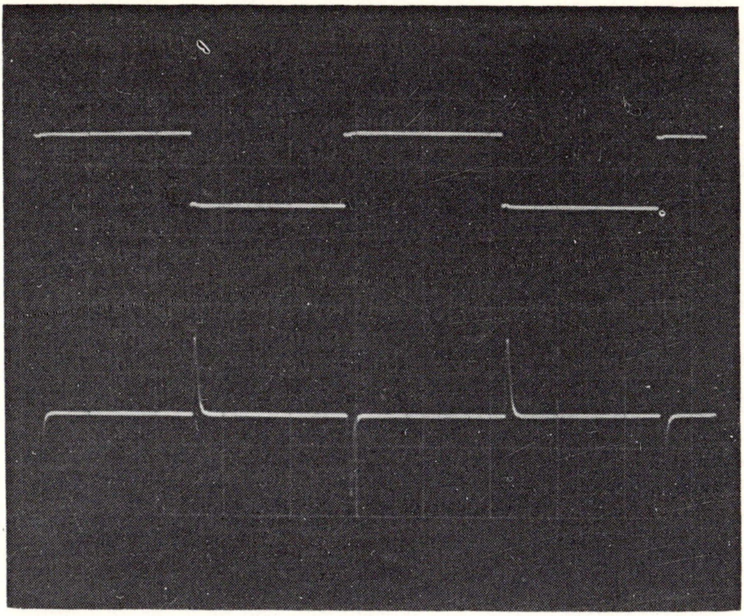

Fig. 16-22. Input and output of differentiator—1 kHz input. Upper trace: Input 1V per cm. Lower trace: Output 10V per cm. Time base: 0.2 msec per cm.

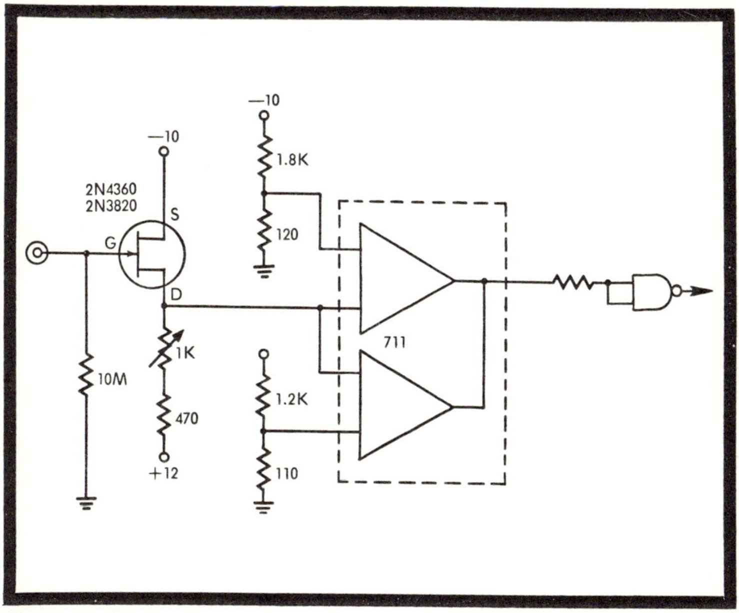

Fig. 16-23. Comparator using 711. This input shaper will accept inputs as low as 15 mV and deliver logic-level output pulses for a counter.

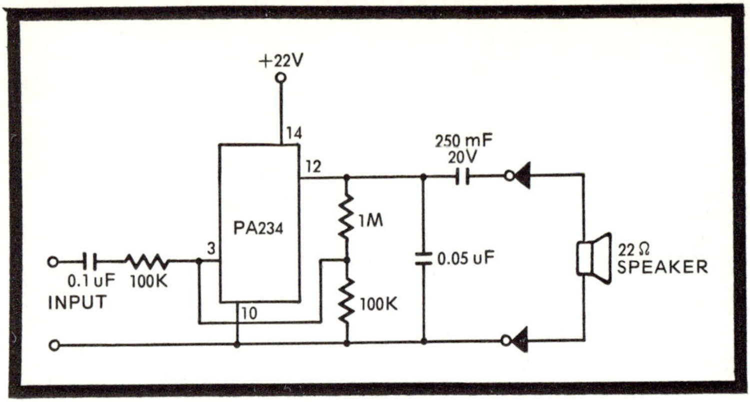

Fig. 16-24. The 1W audio amplifier using PA234.

DIFFERENTIATORS

A differentiator performs a mathematical function, as does the integrator. The output of the differentiator is a function of the rate of change of the input. If the input remains at some dc voltage level, the output will be zero. If the input changes (in this case, sharply from one dc level to another), the output will be proportional to this change, thus generating a pulse such as the one shown Fig. 16-22.

The resulting output from the circuit of Fig. 16-21 is the differentiation of a square wave giving pulses at the output of

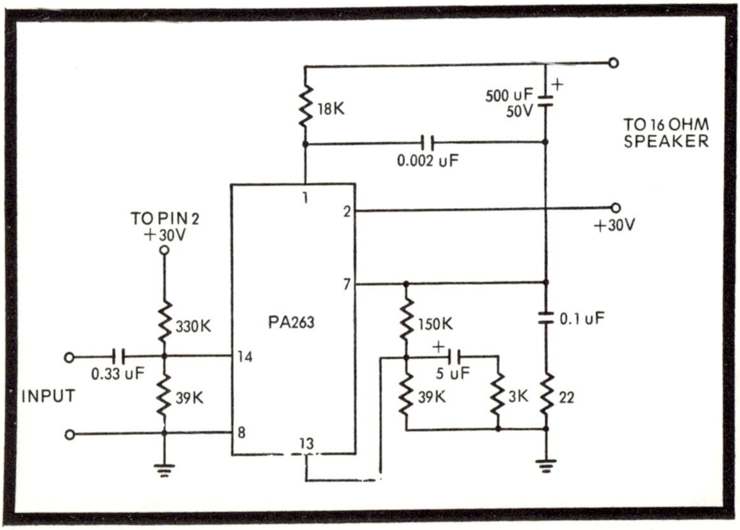

Fig. 16-25. The 3.5W audio amplifier using PA263.

a polarity opposite to that of the change. If the input goes positive, a negative pulse is generated, and vice versa.

With the low cost of IC op amps, it is possible to create some very interesting circuit configurations by cascading various types of circuitry to achieve specific results. Fairchild Semiconductors, and Signetics, among others, have application books with hundreds of circuits for these and other devices.

COMPARATORS

Any op amp can be used as a comparator. The circuit is run open-loop. If the inverting input goes slightly negative (remember the small input required for a 28V output swing), the output will go hard positive—if the input goes slightly positive, the output will go hard negative. The result is an output square wave for a very small ac signal input.

In one type of circuit, the switch on the precision ramp generator (Fig. 16-16) can be a FET controlled by the output of a comparator. The comparator input is from the ramp generator. The result is a combination that generates a recurring ramp.

The circuit of Fig. 16-23 makes use of a dual comparator of a special type. The output of this comparator is either a logical "0" (zero volts) or a logical "1" (5V). The input required is very small; thus, the circuit can be operated with signals as low as 15 mV. The single-comparator version of this circuit is the 710.

Chapter 17

Construction Techniques

Because most of the circuits presented are think-pieces, specific cabinet details, front panels, and similar mechanical specifications have been left out to make room for more circuits and circuit details.

It does seem, however, that some general suggestions are in order. Both Radio Shack and other manufacturers or distributors offer a number of simple construction systems that could be used for breadboarding systems. In addition to these, Vector, Vero, and Eico offer similar systems or equipment. For example, Radio Shack offers "Perfbox" and an experimenter's "P-Box" plus Perfboards and push-in terminals that can be used to advantage for circuits with as many as three transistors—perhaps more. The Calectro (GC) catalog also contains dozens of accessories and hundreds of components that can be put to use in the circuits covered in this book.

PC BOARDS

Circuit-board layouts are included for five of the power amplifiers; similar layouts can be made for the other circuits. Buy Buk and Brady supply printed circuit drafting aids. For this application, 3/16 in. black circles and 1/16 in. tape can be used to best advantage. Lay the circuit out exactly twice the desired size, and have the finished artwork photographed at a lithographic printing shop. Ask for a 50 percent negative. (If you are using the artwork in this book, as for a 100 percent negative, as these are 1:1.)

With this negative, a board can be made using Kepro presensitized circuit-board material, following the instructions supplied with this material. The finished board can be cut to size with an Adel "Nibbler" and drilled with an ordinary high-speed carbon drill. If the material is fiberglass, only 100 holes or so car be drilled (one circuit board) before the drill flutes will be worn away, so figure the cost of one drill as part of the cost of the board.

A list of suppliers' names and addresses follows.

SUPPLIER

Radio Shack 2725 W. 7th Street Fort Worth, Texas 76107	Parts and experimenters' material
Vector Electronics Co., Inc. 12460 Gladstone Avenue Sylmar, California 91342	
TAB Book/Kits Blue Ridge Summit, PA 17214	
Vero Electronics Inc. 176 Central Avenue Framingdale, New York 11735	Breadboarding equipment and material
Lafayette Radio Electronics 111 Jericho Turnpike Syosset, New York	Free catalog
Kepro Circuit Systems Inc. 3630 Scarlet Oak Blvd. St. Louis, Missouri 63122	Printed-circuit material, including presensitized board
W. H. Brandy Co. 727 W. Glendale Avenue Milwaukee, Wisconsin 53201	Front panels, labels, and printed-circuit drafting aids

APPENDIX FERRANTI AGENT ADDRESSES

E-LINE TRANSISTOR SPECIFICATIONS
AND OUTLINE DIMENSIONS
SILICON DIODE SPECIFICATIONS
AND OUTLINE DIMENSIONS
TRANSISTOR CHARACTERISTIC CURVES
AND TYPICAL PARAMETERS
SPECIFIC SAFE OPERATION PARAMETERS
FOR TRANSISTORS USED IN
EXAMPLE CIRCUITS

AGENTS' ADDRESSES—FERRANTI DEVICES

The data sheets that follow are included for information. Ferranti agents' addresses are listed below, so their devices should be obtainable from them or from one of their distributors. If difficulty is encountered, most of the circuits can be built using near equivalents selected by comparing specifications in the data sheets.

The ZTX300, for example, is a simple silicon npn transistor with 50 to 300 h_{FE}. If the voltages are about the same, almost any general-purpose npn audio transistor can be used as a substitute for this device.

TRANSISTOR SPECIFICATIONS

The main characteristics of the devices used for the circuits described are given in the tables that follow.
The transistors are divided into categories according to their power-dissipation capability.
Safe Operating Area (SOA) diagrams are shown at the end.

ZTX 300 (BCW 10) Series

TYPICAL CHARACTERISTICS

h_{FE}/I_c

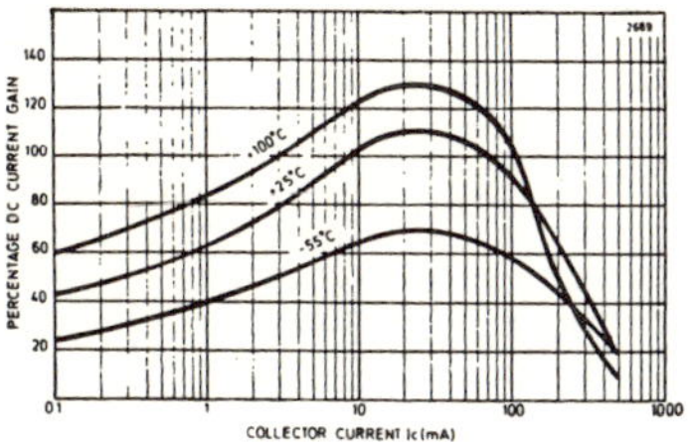

$V_{BE(sat)}$/Amb. temp.

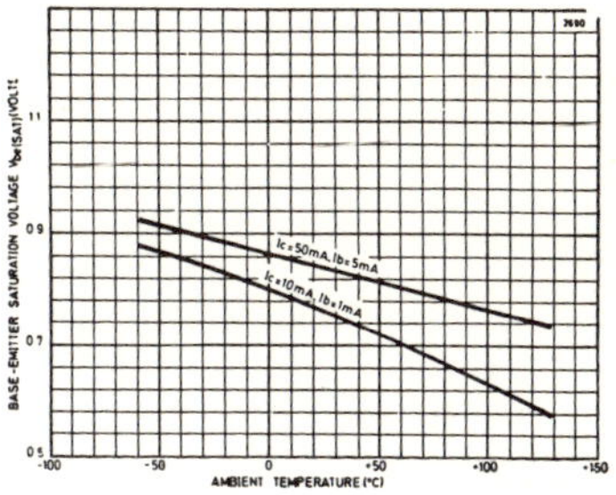

$V_{BE(sat)}/I_c$

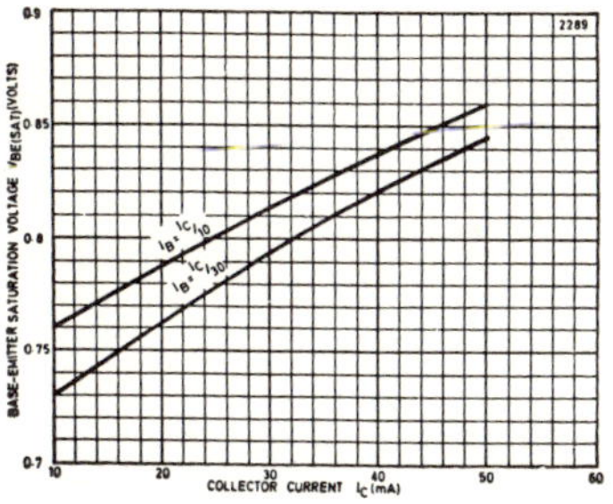

h_{FE}/Amb. temp. at $V_{CE}=6V$

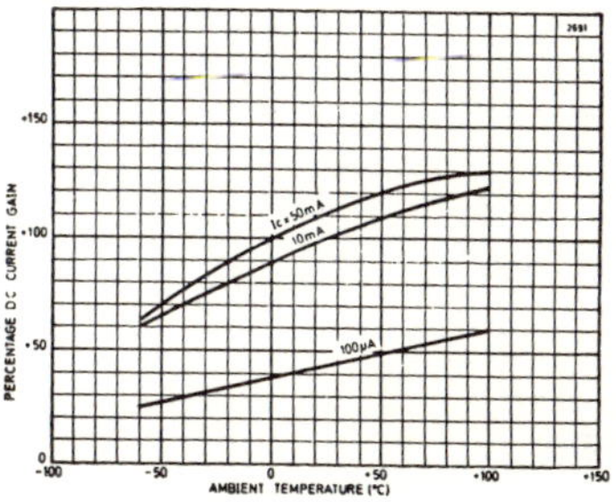

$V_{CE(sat)}/I_c$

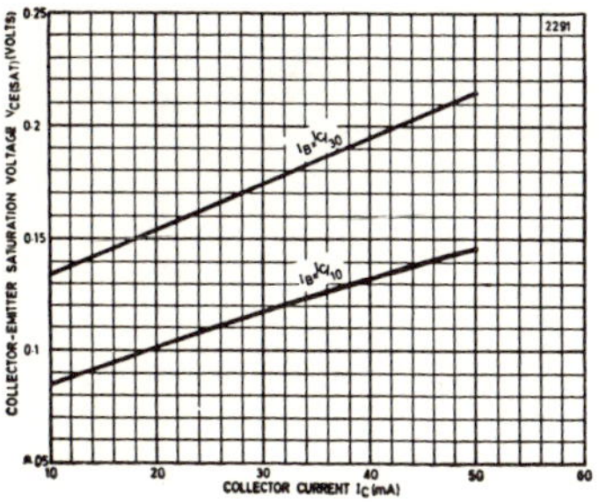

$V_{CE(sat)}$/Amb. temp.

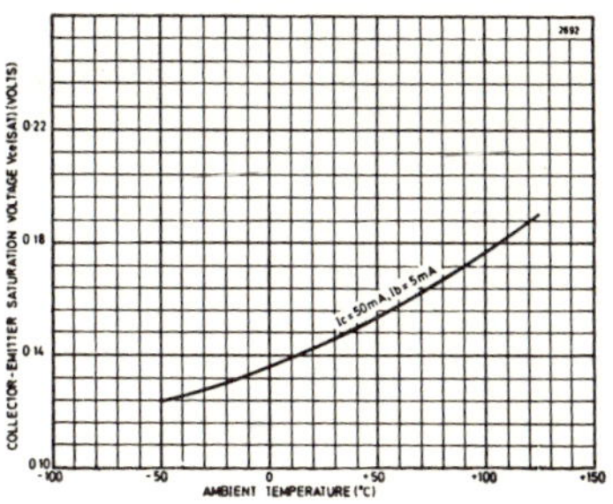

ZTX 310 (BSV23) Series

The ZTX310 Series are n-p-n high speed switching transistors electrically similar to 2N706, BSY95A 2N706A, 2N2369 and 2N2369A.

RATINGS

		ZTX 310 (BSV23)	ZTX 311 (BSV24)	ZTX 312 (BSV25)	ZTX 313 (BSV26)	ZTX 314 (BSV27)	
Max. Collector-Base Voltage	V_{CBO}	25	20	30	40	40	V
Max. Collector-Emitter Voltage	V_{CEO}	12	15	12	15	15	V
Max. Emitter-Base Voltage	V_{EBO}	3	5	5	5	5	V
Max. Peak Collector Current	I_{CM}	500	500	500	500	500	mA
Max. Dissipation (All Types)	P_{tot}						
at $T_{amb}=25°C$		300 mW					
Derating with temp.		Linearly to zero at 125°C					
Max. Operating and Storage Temp. (All Types) ...		−55 to +125°C					

CHARACTERISTICS (At 25°C ambient temperature unless otherwise specified)

Parameter	Symbol	ZTX 310 (BSV23)	ZTX 311 (BSV24)	ZTX 312 (BSV25)	ZTX 313 (BSV26)	ZTX 314 (BSV27)		Conditions
Collector-Base Reverse Current ...	I_{CBO}							
T_{amb} = 25°C Max. ...		200	200	200	200	200	nA	$V_{CB}=15V$ (ZTX310, 311)
T_{amb} = 100°C Max. ...		30	30	30	30	30	μA	$V_{CB}=20V$ (ZTX312, 313, 314)
Collector-Base Breakdown Voltage... Min. ...	BV_{CBO}	25	20	30	40	40	V	$I_C=10\mu A$
Collector-Emitter Sustaining Voltage Min. ...	$V_{CEO(sus)}$	—	15	12	15	15	V	$I_C=10mA$ $I_B=0$

Collector-Emitter Saturation Voltage $V_{CE(sat)}$								
	Max. ...	0·6	—	0·24	0·24	0·2	V	$I_C=10mA$ $I_B=1mA$
	Max. ...	—	—	—	—	0·5	V	$I_C=100mA$ $I_B=10mA$
Base-Emitter Saturation Voltage ... $V_{BE(sat)}$								
	Min. ...	0·75	0·7	0·7	0·7	0·7	V	} $I_C=10mA$ $I_B=1mA$
	Max. ...	0·9	0·9	0·85	0·85	0·85	V	} $I_C=10mA$ $I_B=1mA$
	Max. ...	—	—	—	—	1·1	V	$I_C=30mA$ $I_B=3mA$
	Max. ...	—	—	—	—	1·6	V	$I_C=100mA$ $I_B=10mA$
D.C. Current Gain ... ... h_{FE}								
	Min. ...	20	—	40	40	40		} $I_C=10mA$ $V_{CE}=1V$
	Max. ...	—	—	—	120	120		} $I_C=10mA$ $V_{CE}=1V$
	Min. ...	—	—	35	—	30		$I_C=30mA$ $V_{CE}=1V$
	Min. ...	—	—	—	15	20		$I_C=100mA$ $V_{CE}=1V$
	Min. ...	—	—	20	—	—		$I_C=1mA$ $V_{CE}=6V$
	Min. ...	—	30	—	—	—		$I_C=1mA$ $V_{CE}=0·35V$
	Min. ...	—	50	—	—	40		} $I_C=10mA$ $V_{CE}=0·35V$
	Max. ...	—	200	—	—	120		} $I_C=10mA$ $V_{CE}=0·35V$
T_{amb} = –55°C	Min. ...	—	—	20	—	—		$I_C=10mA$ $V_{CE}=0·35V$
Gain-Bandwidth Product ... ... f_T	Min. ...	200	200	400	500	500	MHz	$I_C=10mA$ $V_{CE}=10V$ f=100mHz
Output Capacitance .. .. c_{ob}	Max. ...	6	6	4	4	4	pF	$V_{BE}=5V$ $I_E=0$ f=1MHz
Storage Time ... ... t_s	Max. ...	60	25	13	13	13	nS	$I_C=I_{B1}=I_{B2}=10mA$
Turn-On Time ... ... t_{on}	Max. ...	—	—	15	12	12	nS	$I_C=10mA$ $I_{B1}=3mA$
Turn-Off Time ... ... t_{off}	Max. ...	—	–	20	18	18	nS	$I_C=10mA$ $I_{B1}=3mA$ $I_{B2}=1·5mA$

ZTX 310 (BSV23) Series

TYPICAL CHARACTERISTICS

ZTX312, 313, 314

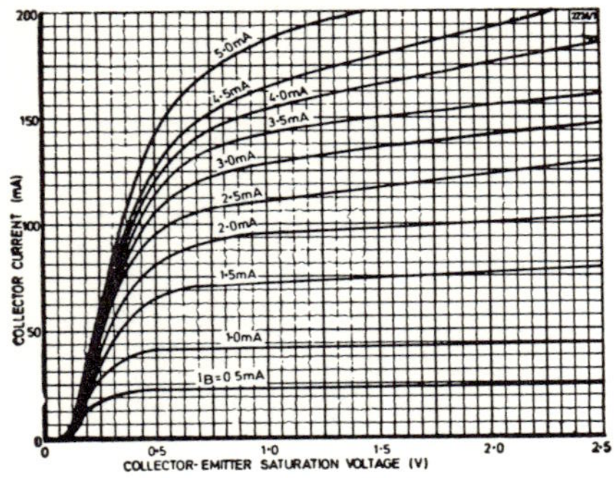

ZTX311, 313, 314

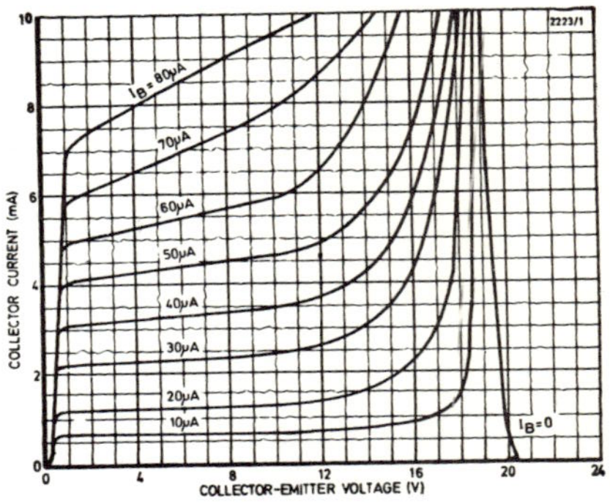

ZTX311, 312, 313, 314

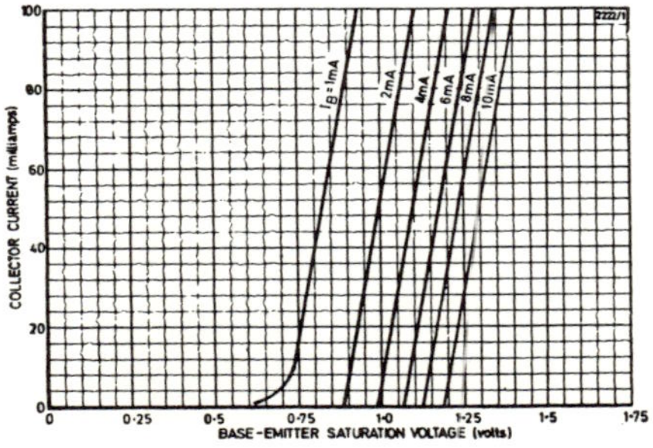

ZTX312

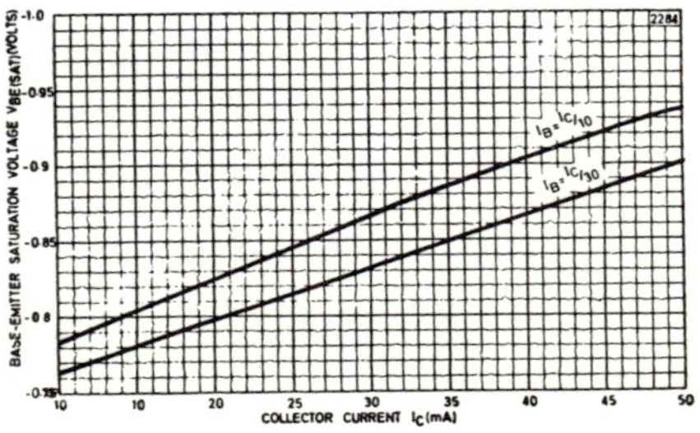

ZTX 310 (BSV23) Series

TYPICAL CHARACTERISTICS

ZTX312

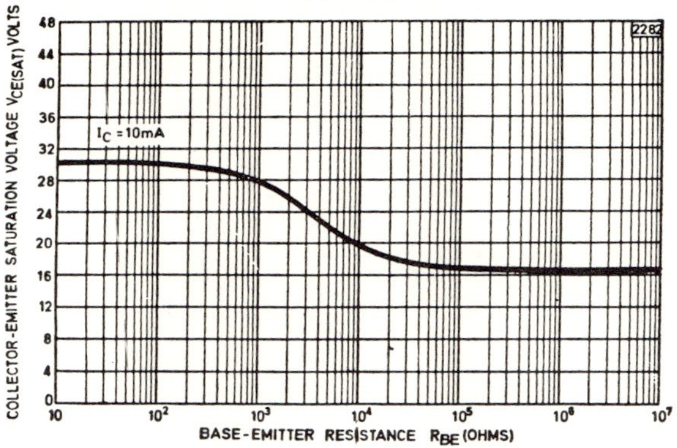

ZTX312

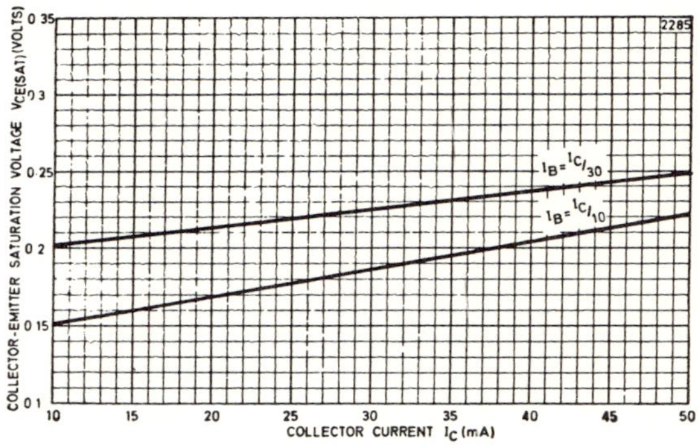

ZTX312

ZTX 320 (BFW 97) Series

The ZTX320 transistor is an n-p-n V.H.F./U.H.F. transistor similar to 2N918.

RATINGS

	Symbol	ZTX320 (BFW97)	ZTX321	Unit
Max. Collector-Base Voltage	V_{CBO}	30	30	V
Max Collector-Emitter Voltage	V_{CEO}	15	15	V
Max. Emitter-Base Voltage	V_{EBO}	3	3	V
Max. Dissipation	P_{tot}			
at $T_{amb}=25°C$		250	250	mW
Derating with temp.		Linearly to zero at 125°C		
Max. Operating and Storage Temp.		−55 to +125		°C

CHARACTERISTICS (At 25°C ambient temperature unless otherwise specified)

Parameter	Symbol	ZTX320 (BFW97)	ZTX321	Unit	Conditions
Max. Collector-Base Reverse Current	I_{CBO}	0·2	0·2	μA	$V_{CB}=30V$, $I_E=0$
Max. Emitter-Base Reverse Current	I_{EBO}	0·2	0·2	μA	$V_{EB}=2V$, $I_C=0$
Max. Collector-Emitter Saturation Voltage	$V_{CE(sat)}$	0·4	—	V	$I_C=10mA$, $I_B=1mA$
		—	0·4	V	$I_C=3mA$, $I_B=0{\cdot}3mA$
Max. Base-Emitter Saturation Voltage	$V_{BE(sat)}$	1·0	—	V	$I_C=10mA$, $I_B=1mA$
		—	1·0	V	$I_C=3mA$, $I_B=0{\cdot}3mA$
Min. D.C. Current Gain	h_{FE}	20	20		$I_C=3mA$, $V_{CE}=1V$
Gain-Bandwidth Product	f_T	600	600	MHz	$I_C=4mA$, $V_{CE}=10V$, f = 100MHz
Max. Input Capacitance	c_{ib}	1·6	1·6	pF	$V_{EB}=0{\cdot}5V$, f = 1MHz
Max. Collector Capacitance	c_{ob}	3	3	pF	$V_{CB}=0$, f = 1MHz
		1·7	1·7	pF	$V_{CB}=10V$, f = 1MHz
Max. Noise Figure	N	6	6	dB	$I_C=1mA$, $V_{CE}=6V$, $R_s=400\Omega$ f = 60MHz
Typical Power Gain	g_{pe}	15	15	dB	$I_C=6mA$, $V_{C3}=12V$, f = 200MHz

ZTX 325, ZTX 326

U.H.F. n-p-n Transistors

Absolute Maximum Ratings	Symbol	ZTX325	ZTX326	Unit
Collector-Base Voltage	V_{CBO}	30	25	V
Collector-Emitter Voltage	V_{CEO}	15	12	V
Emitter-Base Voltage	V_{EBO}	2·5	2·5	V
Collector Current	I_{CM}	50	50	mA
Mean Collector Current ... ... (Averaged over 100μS period)...	I_{AV}	25	25	mA
Dissipation (Ambient Temp.=25°C)	P_{tot}	200	200	mW
Operating Temp. Range	T_{OPT}	−65 to +175		°C
Junction Temperature	T_J	175	175	°C

ZTX 327

VHF-UHF n-p-n transistor

RATINGS (Absolute max. values)

Rating	Symbol	Value	Unit
Collector-Base Voltage	V_{CBO}	55	volts
Collector-Emitter Voltage	V_{CEO}	30	
($R_{BE}=10\Omega$)	V_{CER}	55	volts
Emitter-Base Voltage	V_{EBO}	3·5	volts
Collector Current	I_C	0·4	amp.
Dissipation (at Case Temp. −65 to +25°C)	P_{tot}	500	mW
Dissipation Derating (above 25°C)		3·33	mW/°C
Junction Temperature	T_{jm}	175	°C
Storage Temperature Range		−65 to +175	°C

CHARACTERISTICS (At 25°C unless otherwise stated)

Characteristic	Symbol	Min.	Typ.	Max.	Unit	Test Conditions
Collector Cut-off Current	I_{CEO}	—	—	20	μA	$V_{CE}=28V$, $I_B=0$
Collector-Base Breakdown Voltage	BV_{CBO}	55	—	—	V	$I_E=0$, $I_C=100\mu A$
Collector-Emitter Sustaining Voltage	$V_{CER(sus)}$	55	—	—	V	$I_C=5mA$, $R_{BE}=10$ ohms
	$V_{CEO(sus)}$	30	—	—	V	$I_C=5mA$, $I_B=0$
Collector-Emitter Saturation Voltage	$V_{CE(sat)}$	—	—	1·0	V	$I_B=20mA$, $I_C=100mA$
Collector-Base Capacitance	C_{ob}	—	—	3·0	pF	$V_{CB}=30V$, $I_E=0$, $f=1MHz$
Power Output as Oscillator	P_{out}	200	—	—	mW	$V_{CB}=21V$, $I_C=25mA$ f=600MHz
Gain-Bandwidth Product	f_T	—	800	—	MHz	$V_{CE}=15V$, $I_C=25mA$

CHARACTERISTICS (at 25°C junction temperature unless otherwise specified)

Parameter	Symbol	ZTX325 Min.	ZTX325 Max.	ZTX326 Min.	ZTX326 Max.	Unit	Test Conditions
Collector-Base Current	I_{CBO}	—	10	—	—	nA	$V_{CB}=15V$, $I_E=0$
		—	—	—	1·0	μA	
Collector-Emitter Current (Base short-circuited to emitter)	I_{CES}	—	10	—	—		$V_{CE}=15V$, $V_{BE}=0$
		—	—	—	10	μA	$V_{CE}=12V$, $V_{BE}=0$
Collector-Emitter Sustaining Voltage	$V_{CEO(sus)}$	15	—	12	—	V	$I_C=10mA$, $I_B=0$
D.C. Current Gain	h_{FE}	25	150	25	150		$V_{CE}=1 \cdot 0V$, $I_C=2 \cdot 0mA$
		20	125	20	125		$V_{CE}=1 \cdot 0V$, $I_C=25mA$
Gain-Bandwidth Product	f_T	1·0	—	1·0	—	GHz	$V_{CE}=5 \cdot 0V$, $I_C=2 \cdot 0mA$
		1·3	—	1·0	—	GHz	$V_{CE}=5 \cdot 0V$, $I_C=25mA$
Capacitance, collector depletion layer	C_{Tc}	—	1·5	—	1·5	pF	$V_{CB}=10V$, $I_E=I_e=0$, $f=1 \cdot 0MHz$
Capacitance, emitter depletion layer	C_{Te}	—	2·0	—	2·0	pF	$V_{EB}=0 \cdot 5V$, $I_C=I_c=0$, $f=1 \cdot 0MHz$
Noise Factor	N	—	5·0	—	6·0	dB	$f=500MHz$, $R_S=50\Omega$ $V_{CE}=5 \cdot 0V$, $I_C=2 \cdot 0mA$
Intermodulation Distortion	d_{im}	−53 Typical		—	—	dB	$V_{CE}=6 \cdot 0V$, $I_C=14mA$, $f=217MHz$; $V_O=100mV$, $R_L=37 \cdot 5\Omega$; $f_1=183MHz$, $f_2=200MHz$
‡Output Power (At $T_{case}=25°C$)	P_o	175	—	—	—	mW	$V_{CE}=13 \cdot 5V$, $I_C=22 \cdot 5mA$; $P_{in}=25mW$, $f=500MHz$

‡It is essential that care is taken to reduce steady state current when no h.f. signal is applied.

ZTX 330 (BCW 20), 331 (BCW 22)

These transistors are intended for low level, low noise amplifier applications and are n-p-n complements to ZTX530, ZTX531. They are electrically similar to Types 2N3707 and 2N929 respectively.

RATINGS

	Symbol	ZTX330 (BCW20)	ZTX331 (BCW22)
Max. Collector-Base Voltage	V_{CBO}	30	45 V
Max. Collector-Emitter Voltage	V_{CEO}	30	45 V
Max. Emitter-Base Voltage	V_{EBO}	5	5 V
Max. Peak Collector Current	I_{CM}	500	500 mA
Max. Dissipation at 25°C (both types)	P_{tot}	250 mW	
Derating with Temp.		Linearly to zero at 125°C	
Max. Operating and Storage Temp. (Both types) ...		-55 to +125°C	

CHARACTERISTICS (at 25°C unless otherwise specified)

Parameter	Symbol	ZTX330 (BCW20)	ZTX331 (BCW22)	Conditions
Max. Collector-Base Reverse Current	I_{CBO}	0·2	0·2 μA	$V_{CB}=V_{CO(M)}$, $I_E=0$
Max. Emitter-Base Reverse Current	I_{EBO}	0·2	0·2 μA	$V_{EB}=4V$, $I_E=0$
Min. Collector-Emitter Sustaining Voltage	$V_{CEO(sus)}$	30	45 V	$I_C=5mA$, $I_B=0$
D.C. Current Gain	h_{FE}	100–400		$V_{CE}=5V$, $I_C=100\mu A$
			40–120	$V_{CE}=5V$, $I_C=10\mu A$
Max. Collector-Emitter Saturation Voltage	$V_{CE(sat)}$	0·7	0·7 V	$I_C=10mA$, $I_B=0{\cdot}5mA$
Base-Emitter Saturation Voltage	$V_{BE(sat)}$	0·5–1·0	0·5–1·0 V	$V_{CE}=5V$, $I_C=1mA$
A.C. Current Gain	h_{fe}	60–400		$I_C=100\mu A$ } $V_{CE}=5V$, $f=1kHz$
			60 (min)	$I_C=1mA$ } $V_{CE}=5V$, $f=1kHz$
Max. Collector Capacitance	C_{ob}	8	8 pF	$V_{CB}=5V$, $I_E=0$, $f=1MHz$
Min. Gain-Bandwidth Product	f_T	30	30 MHz	$V_{CB}=5V$, $I_C=500\mu A$, $f=30MHz$
Typical Noise Figure	N	3	3 dB	$I_C=100\mu A$, $V_{CE}=5V$, $R_S=10k\Omega$, $f=1kHz$

ZTX 341 (BSV 28) Series

Nixie Driver Transistors

RATINGS

	Symbol	ZTX 341 (BSV28)	ZTX 342 (BSV29)
Max. Collector-Base Voltage	V_{CBO}	100	120 V
Max. Collector-Emitter Voltage ... ... $I_C=10\mu A$, $R_{BE}=50k\Omega$	V_{CER}	100	120 V
Max. Emitter-Base Voltage	V_{EBO}	5	5 V
Max. Peak Collector Current	I_{CM}	100	100 mA
Max. Dissipation (Both Types) ... ... at $T_{amb}=25°C$	P_{tot}	300 mW	
Derating with temp.		Linearly to zero at 125°C	
Max. Operating and Storage Temp. (Both Types) ...		-55 to +125°C	

CHARACTERISTICS (At 25°C ambient temperature unless otherwise specified)

Parameter	Symbol	ZTX 341 (BSV28)	ZTX 342 (BSV29)	Conditions	
Max. Collector-Emitter Reverse Current	I_{CER}	0·5	— μA	$V_{CE}=80V$	$R_{BE}=50k\Omega$
		—	0·5 μA	$V_{CE}=100V$	
$T_{amb}=100°C$		10	— μA	$V_{CE}=80V$	
		—	10 μA	$V_{CE}=100V$	
Max. Collector-Emitter Saturation Voltage	$V_{CE(sat)}$	0·5	0·5 V	$I_C=2mA$, $I_B=0{\cdot}1mA$	
Max. Base-Emitter Saturation Voltage	$V_{BE(sat)}$	1·0	1·0 V		
Min. D.C. Current Gain	h_{FE}	30	30	$I_C=2mA$	$V_{CE}=1V$
		—	40	$I_C=10mA$	

ZTX 350 Series

Single p-channel enhancement mode MNOS transistor with integral gate protection, source and substrate common.

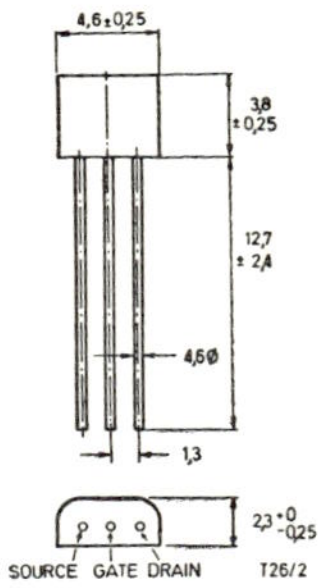

Dimensions in millimetres

MAXIMUM RATINGS	Symbol	Rating	Unit
Drain-Source Voltage	V_{DS}	–20	V
Gate-Substrate Voltage		–20	V
Dissipation at $T_{amb} = 25°C$	P_{tot}	300	mW
Derating with temp.		Linearly to zero	
Storage Temperature Range	T_{STG}	–55 to +125	°C
Operating Temp. Range	T_{OPT}	0 to +100	°C

CHARACTERISTICS (At 25°C ambient temperature unless otherwise specified)

Parameter	Symbol	Min.	Typ.	Max.	Unit	Conditions
Gate-Input Leakage Current	I_{GSS}	—	—	3	nA	$V_{GS} = -20V$, $V_{DS} = 0$
Drain-Source Leakage Current	I_{DSS}	—	—	3	nA	$V_{DS} = -20V$, $V_{GS} = 0$
Threshold Voltage	V_{TH}	1·0	-2·0	6·0	V	$V_{DS} = -20V$ $I_{DS} = -10\mu A$
Gate-Source Voltage	V_{GS}	—	1·3	—	V	$V_{DS} = -20V$, $I_{DS} = -10mA$
Mutual Conductance	g_m	0·5	1·3	2·5	mA/V	$V_{GS} = V_{DS} = -10V$
Drain-Source ON Resistance	R_{on}	—	700	1500	Ω	$I_{DS} = -1mA$, $V_{GS} = -10V$
Gate to Source and Drain Capacitance	C_{GDS}	—	—	4·5	pF	$V_{GS} = 0$, f = 1MHz
Drain-Source Capacitance	C_{DS}	—	1·5	—	pF	$V_{DS} = 0$, f = 1MHz
Drain-Gate Capacitance	C_{DG}	—	1·5	—	pF	$V_{DG} = 0$, f = 1MHz

ZTX 360

High speed n-p-n, core driving transistor

RATINGS (Absolute max. values)

	Symbol		Unit
Collector-Base Voltage	V_{CBO}	60	volts
Collector-Emitter Voltage	V_{CEO}	40	volts
Emitter-Base Voltage	V_{EBO}	5	volts
Peak Collector Current	I_C	1·0	Amp
Dissipation (at Ambient Temp. = 25°C)...	P_{tot}	500	mW
Operating Temperature Range	T_{OPT}	-65 to +175	°C

CHARACTERISTICS
(At 25°C junction temperature unless otherwise stated)

Parameter	Symbol	Min.	Max.	Unit	Test Conditions
Collector-Base Reverse Current ...	I_{CBO}				
At T_j = 25°C		—	500	nA	$V_{CB} = 40$, $I_E = 0$
At T_j = 150°C		—	300	μA	$V_{CB} = 40$, $I_E = 0$
Collector-Emitter Sustaining Voltage ...	$V_{CEO(sus)}$	—	40	volt	$I_C = 10mA$, $I_B = 0$
Collector-Emitter Saturation Voltage...	$V_{CE(sat)}$	—	0·6	volt	$I_C = 500mA$, $I_B = 50mA$
Base-Emitter Saturation Voltage ...	$V_{BE(sat)}$	0·7	1·2	volt	$I_C = 500mA$, $I_B = 50mA$
D.C. Current Gain	h_{FE}	25	150		$I_C = 500mA$, $V_{CE} = 1·0V$
A.C. Current Gain	h_{fe}	2·5	—		$I_C = -50mA$, $V_{CE} = 10V$, $f = 100MHz$
Input Capacitance	C_{ib}	36 (Typ.)		pF	$-V_{BE} = 0·5V$, $I_C = 0$, $f = 1MHz$
		—	50	pF	
Output Capacitance	C_{ob}	5·75 (Typ.)		pF	$V_{CB} = 10V$, $I_E = 0$, $f = 1MHz$
		—	10·0	pF	
Recovered Charge	Q_s	3·8 (Typ.)		nC	$I_C = 500mA$, $I_B = 50mA$, $V_{CC} = 30V$
		—	5·0	nC	
Turn-on Time	t_{on}	—	40	nS	$V_{CC} = 30V$, $I_C = 500mA$, $I_{B(on)} = 50mA$, $-V_{BE(off)} = 2V$
Turn-off Time	t_{off}	—	75	nS	$V_{CC} = 30V$, $I_{B(on)} = -I_{B(off)} = 50mA$

TEST CIRCUIT FOR TURN-ON AND TURN-OFF TIME MEASUREMENT

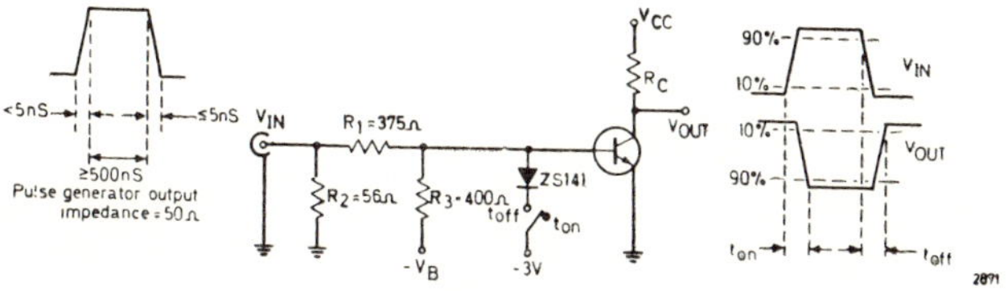

Measurement (R_C = 60 ohms)	$-V_{BE}$ volts	V_{in} volts
t_{on}	4·0	24·75
t_{off}	16·7	37·5

ZTX 382, 383, 384

These are n-p-n transistors designed especially for low noise applications.

RATINGS

	Symbol	ZTX382 382B 382C	ZTX383 383B 383C	ZTX384 384B 384C	Unit
Max. Collector-Base Voltage	V_{CBO}	50	45	45	V
Max. Collector-Emitter Voltage	V_{CEO}	45	30	30	V
Max. Emitter-Base Voltage	V_{EBO}	6	6	6	V
Max. Peak Collector Current	I_{CM}	100	100	100	mA
Max. Dissipation at T_{amb} 25°C	P_{tot}	450	450	450	mW
Derating with Temperature		3	3	3	mW/°C
Max. Junction Temperature	T_{jn}	175	175	175	°C
Storage Temperature Range		–55 to + 175	–55 to + 175	–55 to + 175	°C

CHARACTERISTICS (at 25°C ambient temperature unless otherwise specified)

Parameter	Symbol	ZTX382 Min.	ZTX382 Typ.	ZTX382 Max.	ZTX383 Min.	ZTX383 Typ.	ZTX383 Max.	ZTX384 Min.	ZTX384 Typ.	ZTX384 Max.	Unit	Conditions
Collector-Base Reverse Current	I_{CBO}			15			15			15	nA	V_{CB} = 30V, I_E = 0
Emitter-Base Reverse Current	I_{EBO}						15			15	nA	V_{EB} = 4V, I_C = 0
D.C. Current Gain	h_{FE}	40			40			100				I_C = 10μA, V_{CE} = 5V
		100		850	100		850	250	400			I_C = 2mA, V_{CE} = 5V
		80			80			130				†I_C = 100mA, V_{CE} = 5V

Collector-Emitter Saturation Voltage	$V_{CE(sat)}$			0·25			0·25			0·25	V	I_C = 10mA, I_B = 0·5mA
				0·6			0·6			0·6	V	I_C = 100mA, I_B = 5mA†
Base-Emitter Saturation Voltage	$V_{BE(sat)}$			1·2			1·2			1·2	V	I_C 100mA, I_B = 5mA
A.C. Current Gain	h_{fe}											
ZTX384								240		900		I_C = 2mA, V_{CE} = 5V; f = 1kHz
ZTX384B								240		500		
ZTX384C								450		900		
ZTX383					240		900					
ZTX383B					240		500					
ZTX383C					450		900					
ZTX382		240		900								
ZTX382B		240		500								
ZTX382C		450		900								
Base-Emitter Voltage ..	V_{BE}		0·5			0·5			0·5		V	I_C = 10μA, V_{CE} = 5V
			0·55			0·55			0·55		V	I_C = 100μA, V_{CE} = 5V
		0·55		0·7	0·55		0·7	0·55		0·7	V	I_C = 2mA, V_{CE} = 5V
			0·68			0·68			0·68		V	I_C = 10mA, V_{CE} = 5V
Output Capacitance ..	C_{ob}		4·5			4·5			4·5		pF	f = 1MHz
Collector-Base Capacitance ..	C_{cb}			←		$C_{cb} \simeq C_{ob}$		→				V_{CB} = 5V, I_E = 0
Gain-Bandwidth Product ..	f_T	150			150			150			MHz	I_C = 10mA, V_{CE} = 5V; f = 100MHz
	h_{fe}	1·5			1·5			1·5				
Noise Figure (See graphs) ..	N			4			4			4	dB	V_{CE} = 5V, I_C = 10μA, f = 1kHz, R_S = 2kΩ, f = Δ10Hz
Equivalent Noise Voltage ..	E_n									*0·135	μV	V_{CE} = 5V, I_C = 200μA; R_S = 2kΩ, f = 10–50Hz
Noise Figure Broad Band ..	$\bar{F}$			6			6			4	dB	V_{CE} = 5V, I_C = 200μA; R_G = 2kΩ, f = 15·7kHz; f1 = 10Hz, f2 = 10kHz

*Derived from the base noise—voltage.
†Measured under pulsed conditions. Pulse width 200μS, duty cycle ≤ 2%.

TYPICAL CHARACTERISTICS

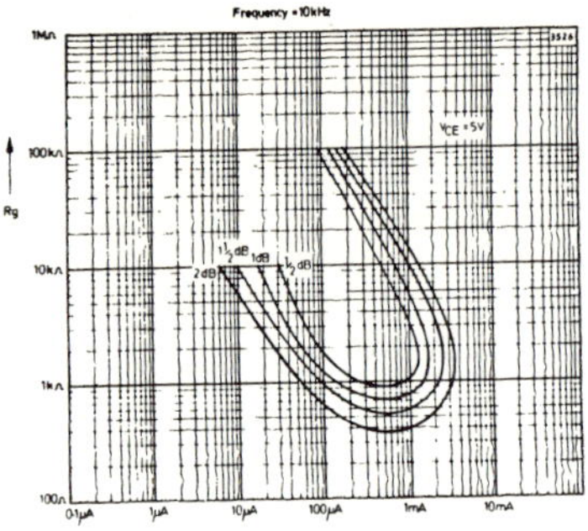

Collector Current

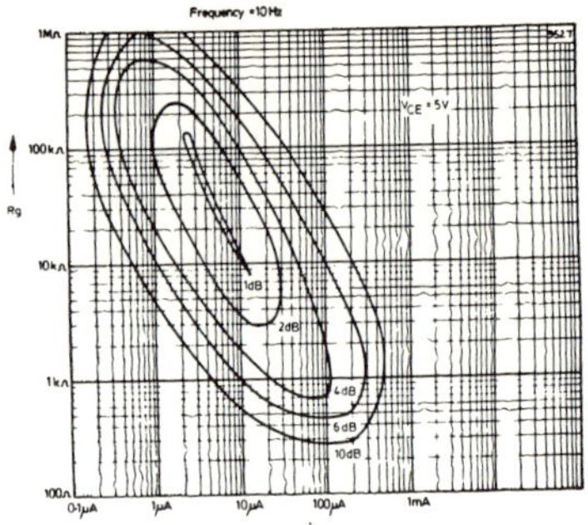

Collector Current

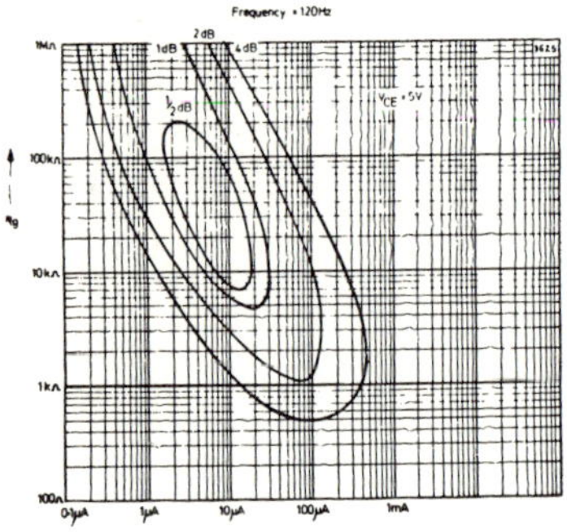

Collector Current

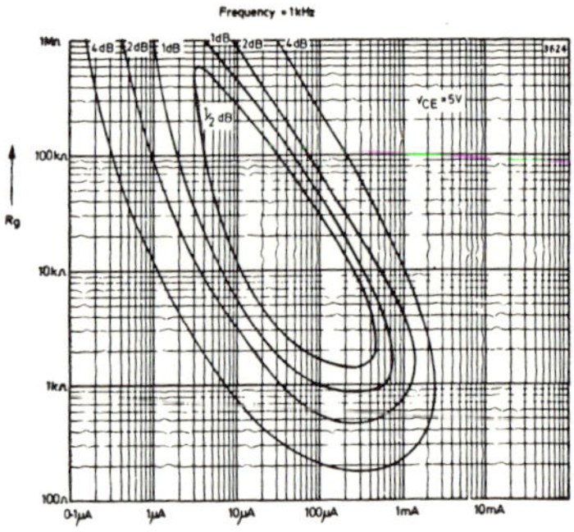

Collector Current

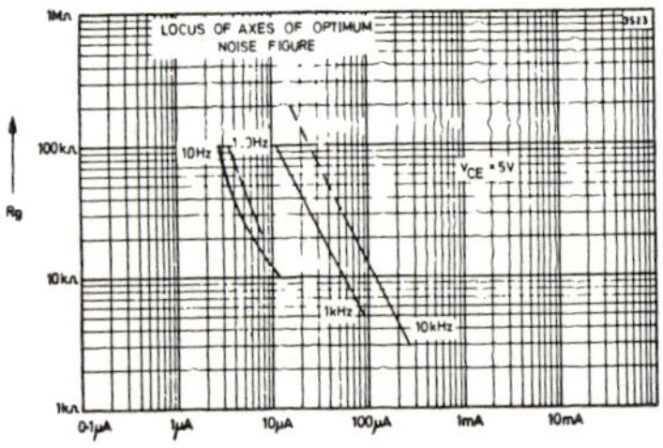

Collector Current

TYPICAL CHARACTERISTICS

ZTX502
h_{FE}/I_C

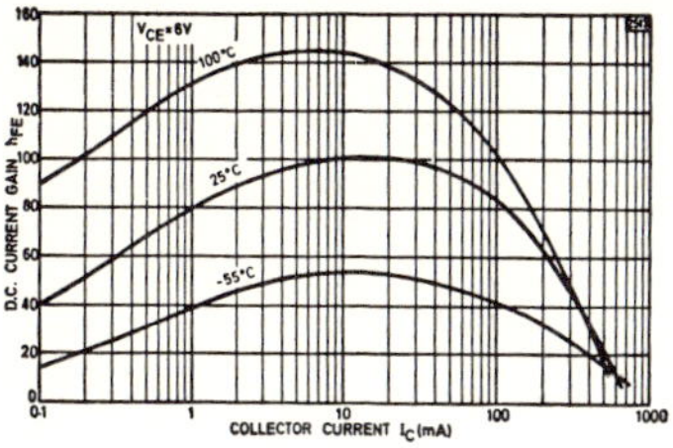

ZTX502
h_{FE}/Amb. Temp. at $V_{CE}=-6V$

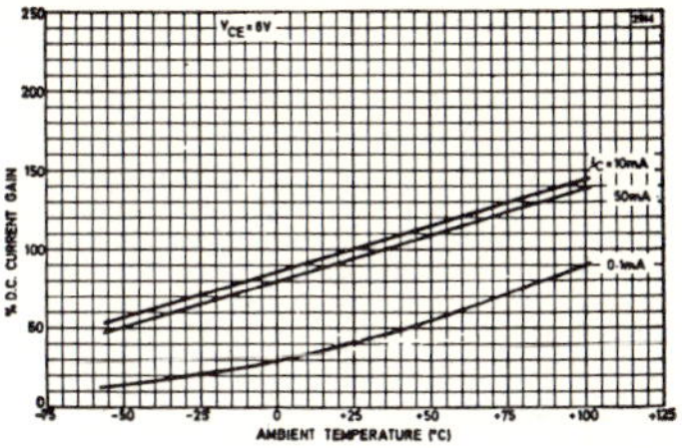

ZTX502
$V_{BE(sat)}$/Amb. Temp.

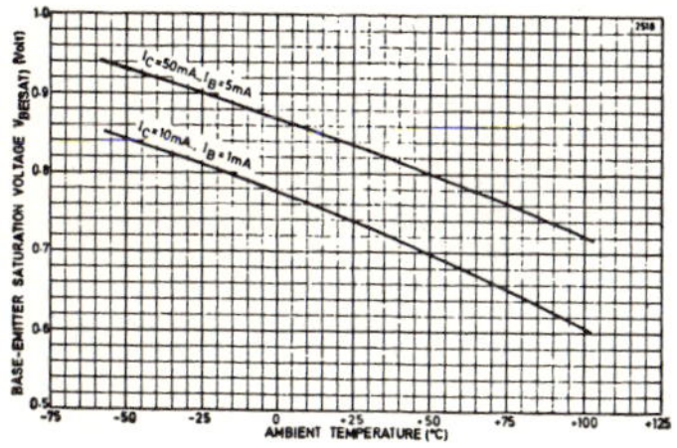

ZTX502
I_C/V_{CE}

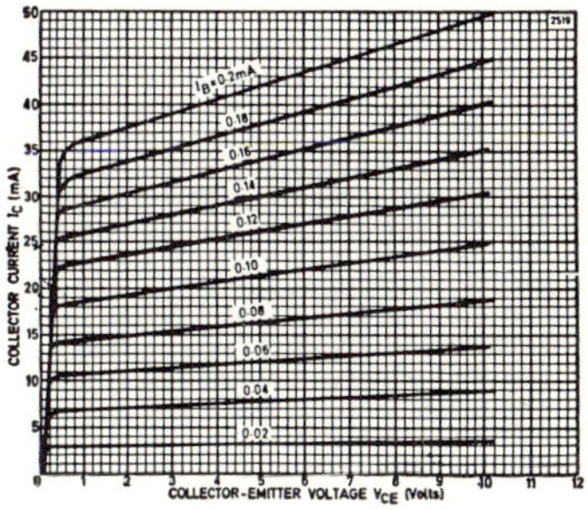

ZTX502
$V_{CE(sat)}/I_C$

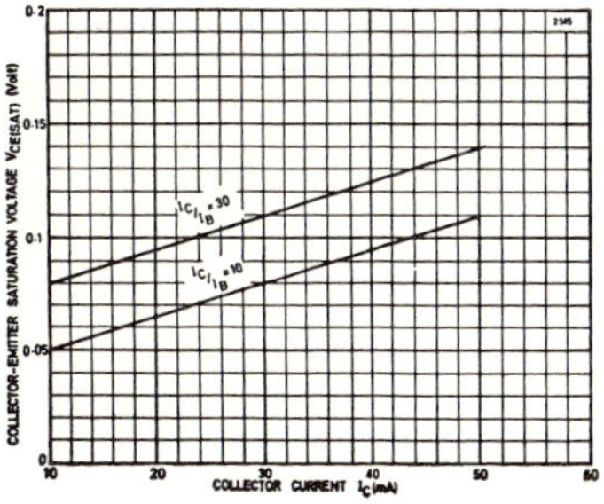

ZTX502
$V_{CE(sat)}$/Amb. Temp.

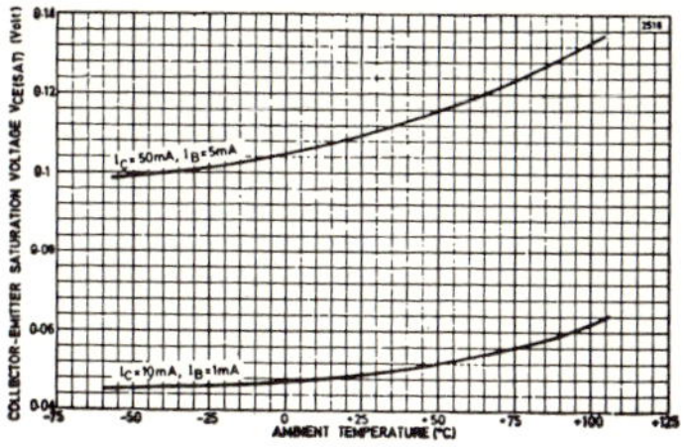

ZTX 500 (BCW 11) Series

These transistors are p-n-p complements to the ZTX300 Series. They are electrically similar to the ZT180 Series.

RATINGS

	Symbol	ZTX 500 (BCW11)	ZTX 501 (BCW13)	ZTX 502 (BCW15)	ZTX 503 (BCW17)	ZTX 504 (BCW19)
Max. Collector-Base Voltage	V_{CBO}	−25	−35	−35	−45	−70 V
Max. Collector-Emitter Voltage	V_{CEO}	−25	−35	−35	−45	−70 V
Max. Emitter-Base Voltage	V_{EBO}	−5	−5	−5	−5	−5 V
Max. Peak Collector Current	I_{CM}	500	500	500	500	500 mA
Max. Dissipation (All types)	P_{tot}					
at T_{amb}=25°C		300 mW				
Derating with temp.		Linearly to zero at 125°C				
Max. Operating and Storage temp. (All types)		−55 to +125°C				

CHARACTERISTICS (At 25°C ambient temperature unless otherwise specified)

Parameter	Symbol	ZTX 500 (BCW11)	ZTX 501 (BCW13)	ZTX 502 (BCW15)	ZTX 503 (BCW17)	ZTX 504 (BCW19)	Conditions
Max. Collector-Base Reverse Current	I_{CBO}	0·2	0·2	0·2	0·2	0·2 μA	$V_{CB}=V_{CBOM}$ $I_E=0$
Max. Emitter-Base Reverse Current	I_{EBO}	0·2	0·2	0·2	0·2	0·2 μA	$V_{EB}=-4V$ $I_C=0$
Min. Collector-Emitter Sustaining Voltage	$V_{CEO(sus)}$	−25	−35	−35	−45	−70 V	$I_C=-5mA$ $I_B=0$
Max. Collector-Emitter Saturation Voltage	$V_{CE(sat)}$	−0·35	−0·25	0·25	−0·35	−0·6 V	$I_C=-50mA$ $I_B=-5mA$
Base-Emitter Saturation Voltage	$V_{BE(sat)}$						
Minimum		0·65	0·65	0·65	0·65	0·65V	$I_C=-10mA$ $I_B=-1mA$
Maximum		1·0	1·0	1·0	1·0	1·0 V	$I_C=-10mA$ $I_B=-1mA$
D.C. Current Gain	h_{FE}						
Minimum		—	—	20	—	—	$V_{CE}=-6V$ $I_C=-100\mu A$
Minimum		50	50	100	50	50	$V_{CE}=-6V$ $I_C=-10mA$
Maximum		300	300	300	300	300	$V_{CE}=-6V$ $I_C=-10mA$
Minimum		—	—	50	—	—	$V_{CE}=-6V$ $I_C=-50mA$
Min. Gain-Bandwidth Product...	f_T	150	150	150	150	150 MHz	$V_{CE}=-6V$ $I_C=-10mA$ f=100MHz
Max. Output Capacitance ...	C_{ob}	6	6	6	6	6 pF	$V_{CB}=-6V$ $I_E=0$ f=1MHz
Typ. Noise Figure	N	7	7	7	7	7 dB	f=1kHz $R_s=1500\Omega$ $I_C=-100\mu A$

ZTX 510 (BSV 33)

This transistor is a **p-n-p** device electrically identical to type TIS50 and electrically similar to type 2N2894.

RATINGS

Max. Collector-Base Voltage	V_{CBO}	−12 V
Max. Collector-Emitter Voltage	V_{CEO}	−12 V
Max. Emitter-Base Voltage	V_{EBO}	−4 V
Max. Peak Collector Current	I_{CM}	−200 mA
Max. Dissipation at 25°C	P_{tot}	250 mW
Max. Operating and Storage Temp.		−55°C to +125°C

CHARACTERISTICS (at 25°C unless otherwise stated)

				Conditions	
Max. Collector-Base Reverse Current	I_{CBO}	100	nA	$V_{CB}=-6V$	
Min. Collector-Emitter Sustaining Voltage	$V_{CEO(sus)}$	−12	V	$I_C=-10mA$	
D.C. Current Gain	h_{FE}	30 min.		$V_{CE}=-0{\cdot}3V$	*$I_C=-10mA$
		40–150		$V_{CE}=-0{\cdot}5V$	*$I_C=-30mA$
		20 min.		$V_{CE}=-1V$	*$I_C=-100mA$
Max. Collector-Emitter Saturation Voltage	$V_{CE(sat)}$	−0·15	V	*$I_C=-10mA$	$I_B=-1mA$
		−0·2	V	*$I_C=-30mA$	$I_B=-3mA$
		−0·5	V	*$I_C=-100mA$	$I_B=-10mA$
Base-Emitter Saturation Voltage	$V_{BE(sat)}$	0·76–0·98	V	*$I_C=-10mA$	$I_B=-1mA$
		0·82–1·2	V	*$I_C=-30mA$	$I_B=-3mA$
		1·7	V	*$I_C=-100mA$	$I_B=-10mA$
Min. Gain Bandwidth Product	f_T	400	MHz	$V_{CB}=-5V$	$I_C=-30mA$, f 100MHz
Max. Output Capacitance	C_{ob}	6	pF	$V_{CB}=-5V$	f=140KHz
Max. Input Capacitance	C_{ib}	6	pF	$V_{EB}=-0{\cdot}5V$	f=140KHz
Switch-On Time	t_{on}	60	nS	$I_O=30mA$	$I_{B1}=I_{B2}=1{\cdot}5mA$
Switch-Off Time	t_{off}	90	nS		

*Pulsed—Pulse duration ⩽ μSec. Duty cycle ⩽2%.

ZTX 530 (BCW 21), 531 (BCW 23)

These transistors are intended for low level, low noise amplifier applications and are p-n-p complements to ZTX330 and ZTX331. They are electrically similar to types 2N4058 and 2N2064 respectively.

RATINGS

		ZTX530 (BCW21)	ZTX531 (BCW23)
Max. Collector-Base Voltage	V_{CBO}	−30	−45 V
Max. Collector-Emitter Voltage	V_{CEO}	−30	−45 V
Max. Emitter-Base Voltage...	V_{EBO}	−5	−5 V
Max. Peak Collector Current	I_{CM}	500mA	
Max. Dissipation at 25°C (Both types)	P_{tot}	250mW	
Derating with Temp.		linearly to zero at +125°C	
Max. Operating and Storage Temp. (Both types) ...		−55°C to +125°C	

CHARACTERISTICS (at 25°C unless otherwise stated)				Conditions
Max. Collector-Base Reverse Current	I_{CBO}	−0·2	−0·2 μA	$V_{CB}=V_{CB(max)}$, $I_E=0$
Max. Emitter-Base Reverse Current	I_{EBO}	−0·2	−0·2 μA	$V_{EB}=-4V$, $I_E=0$
Min. Collector-Emitter Sustaining Voltage	$V_{CEO(sus)}$	−30	−45 V	$I_C=-5mA$, $I_E=0$
D.C. Current Gain	h_{FE}	100-400		$V_{CE}=-5V$, $I_C=-100\mu A$
			40-120	$V_{CE}=-5V$, $I_C=-10\mu A$
Max. Collector-Emitter Saturation Voltage	$V_{CE(sat)}$	−0·7	−0·7 V	$I_C=-10mA$, $I_B=-0{\cdot}5mA$
Base-Emitter Saturation Voltage	$V_{BE(sat)}$	0·5-1·0	0·5-1·0 V	$V_{CE}=-5V$, $I_C=-1mA$
A.C. Current Gain	h_{fe}	60-400		$I_C=-100\mu A$ } $V_{CE}=-5V$, $f=1kHz$
			60 (min)	$I_C=-1mA$ } $V_{CE}=-5V$, $f=1kHz$
Max. Collector Capacitance	c_{ob}	8	8 pF	$V_{CB}=-5V$, $I_E=0$, $f=1MHz$
Min. Gain-Bandwidth Product	f_T	30	30 MHz	$V_{CB}=-5V$, $I_C=-500\mu A$, $f=30MHz$
Typical Noise Figure	N	3	3 dB	$I_C=-100\mu A$, $V_{CE}=-5V$, $R_S=10k\Omega$, $f=1kHz$

ZDX Series Diode Pairs

These are general purpose high speed planar diode pairs

RATINGS	Symbol	ZDX1F	ZDX2F	ZDX1R	ZDX2R	ZDX3F	ZDX3R	ZDX4F	ZDX4R	Unit
ax. Crest Reverse Voltage	V_{RWM}	15	30	15	30	50	50	100	100	V
ax. Forward Current (per common lead)	I_F	250	250	250	250	250	250	250	250	mA
ax. Dissipation at 25°C	P_{tot}	300	300	300	300	300	300	300	300	mW

ARACTERISTICS (Each diode at 25°C unless otherwise specified)

Parameter	Symbol	ZDX1F	ZDX1R	ZDX2F	ZDX2R	ZDX3F	ZDX3R	ZDX4F	ZDX4R	Unit	Conditions
ax. Forward Voltage Drop	V_F	1·0	1·0	1·0	1·0	—	—	—	—	V	I_F=10mA
		—	—	—	—	1·2	1·2	1·2	1·2		I_F=200mA
ax. Reverse Current	I_R	1·0	1·0	1·0	1·0	1·0	1·0	1·0	1·0	μA	At V_{RWM}
ax. Capacitance	C_{max}	4·0	4·0	4·0	4·0	2·5*	2·5*	2·5*	2·5*	pF	V_R=1V, f=1MHz
ax. Reverse Recovery Time†	t_{rr}	7·0	7·0	7·0	7·0	—	—	—	—	nS	I_F=I_R=50-400mA

*Typical. †Time for I_R to recover to 10% of I_R peak.

Note:—Suffix F (e.g. ZDX1F) denotes common cathode configuration.
Similarly, suffix R denotes common anode configuration.

PHYSICAL DATA
(See E-Line Transistors below)

LEAD CONNECTIONS
(See outline drawing)

LEAD CONFIGURATION
(See E-Line Transistors below)

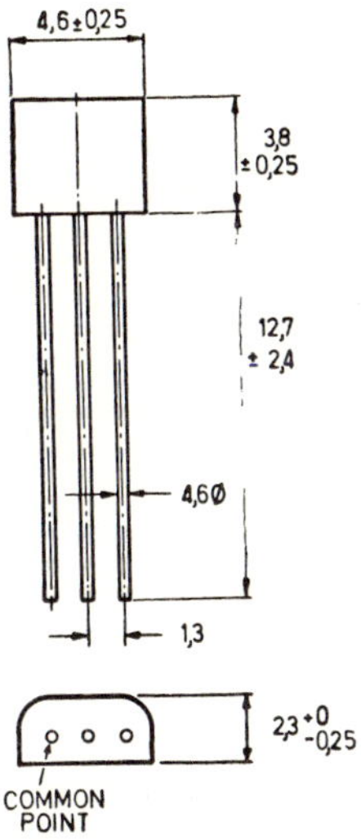

Dimensions in millimetres

E-Line Transistors

PHYSICAL DATA

Max. Body Thickness	...	2,30 mm. (0·090″)
Max. Body Height	...	4,05 mm. (0·160″)
Max. Body Width	...	4,85 mm. (0·190″)
Nom. Lead Length	...	12,7 mm. (0·5″)

LEAD CONNECTIONS

See outline drawing.

LEAD CONFIGURATION

The plastic encapsulation is specifically designed to be compatible with the standard 1,27mm. (0·050″) hole pitch for P.C.B's and the devices are normally supplied with straight leads.

The leads can be preformed, on request, to the TO-5 configuration and when this is done suffix K is added to the type number. Similarly, suffix L indicates that the leads have been preformed to the TO-18 configuration. For flat mounting, suffix M is added to the type number.

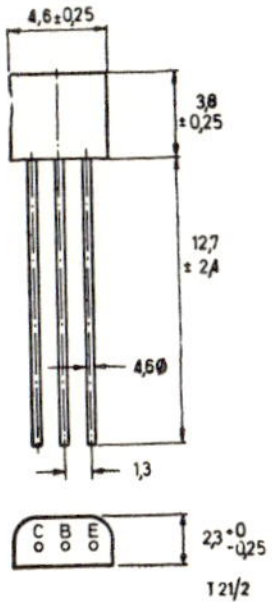

Dimensions in millimetres

BFS 59, 60, 61

n-p-n Transistors complementary to BFS96, 97, 98.

Absolute Maximum Ratings	Symbol	BFS59	BFS60	BFS61	Unit
Collector-Base Voltage	V_{CBO}	60	60	80	V
Collector-Emitter Voltage	V_{CEO}	30	40	60	V
Emitter-Base Voltage	V_{EBO}	5	5	5	V
Peak Collector Current	I_{CM}	1	1	1	A
Dissipation (all types) at $T_{amb}=25°C$	P_{tot}	500	500	500	mW
at $T_{case}=25°C$		2	2	2	W
Typical Thermal Resistance (all types)	θ_j—Amb.		300		°C/W
	θ_j—case		75		°C/W
Operating and Storage Temperature Range (all types)			-55 to +175		°C

CHARACTERISTICS (at 25°C ambient temperature unless otherwise specified)

Parameter	Symbol	BFS59 Min.	BFS59 Max.	BFS60 Min.	BFS60 Max.	BFS61 Min.	BFS61 Max.	Unit	Conditions
Collector-Base Reverse Current	I_{CBO}	—	0·1	—	0·1	—	0·1	μA	$V_{CB}=V_{CEO}$, $I_E=0$
Emitter-Base Reverse Current	I_{EBO}	—	0·1	—	0·1	—	0·1	μA	$V_{EB}=4V$, $I_C=0$
Collector-Emitter Sustaining Voltage	$V_{CEO(sus)}$	30	—	40	—	60	—	V	$I_C=10mA$ (pulsed)
Base-Emitter Saturation Voltage	$V_{BE(sat)}$	—	1·1	—	1·1	—	1·1	V	$I_C=150mA$, $I_B=15mA$
Collector-Emitter Saturation Voltage	$V_{CE(sat)}$	—	0·35	—	0·25	—	0·35	V	$I_C=150mA$, $I_B=15mA$
D.C. Current Gain	h_{FE}	—	—	60	—	25	—	—	$I_C=10mA$, $V_{CE}=10V$ (pulsed)
		40	300	100	300	40	160	—	$I_C=150mA$, $V_{CE}=10V$ (pulsed)
		—	—	15	—	10	—	—	$I_C=1A$, $V_{CE}=10V$ (pulsed)
Gain-Bandwidth Product	f_T	150	—	150	—	150	—	MHz	$I_C=50mA$, $V_{CE}=10V$, $f=100MHz$
Output Capacitance	C_{ob}	—	15	—	15	—	15	pF	$V_{CB}=10V$, $I_E=0$, $f=1MHz$

TYPICAL SMALL SIGNAL 'h' PARAMETERS

	Symbol	BFS59	BFS60	BFS61
Input Impedance	h_{ie}	450	450	450
Reverse Voltage Transfer Ratio	$h_{re} \times 10^{-6}$	150	150	150
A.C. Current Gain	h_{fe}	130	130	130
Output Admittance	h_{oe}	47	47	47

DISSIPATION
$V_{CE} = 10V, I_C = 10mA, f = 1kHz$

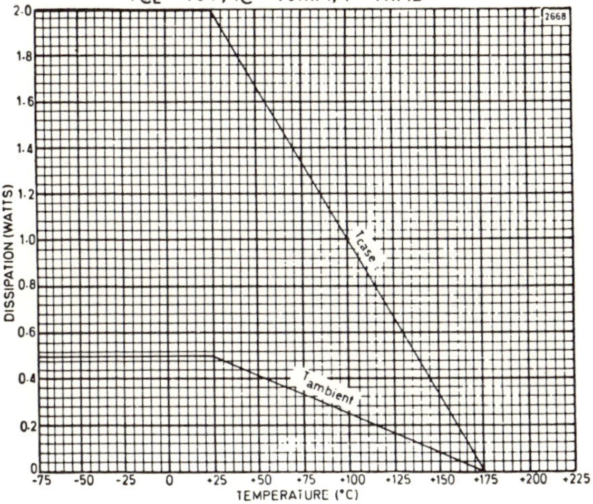

TYPICAL
CHARACTERISTICS
BFS59, 60, 61

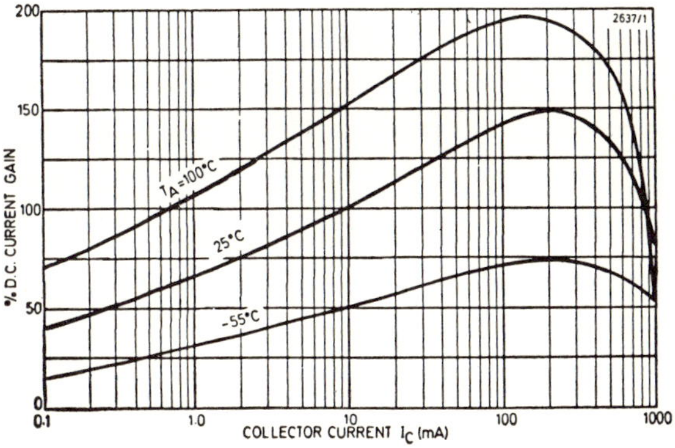

TYPICAL CHARACTERISTICS

BFS59, 60, 61

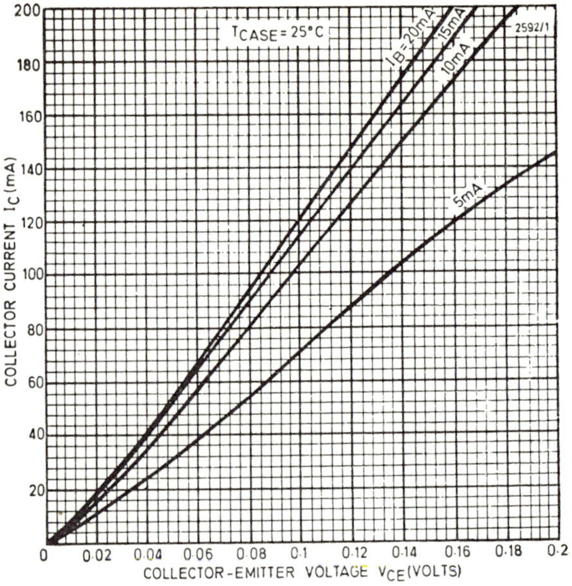

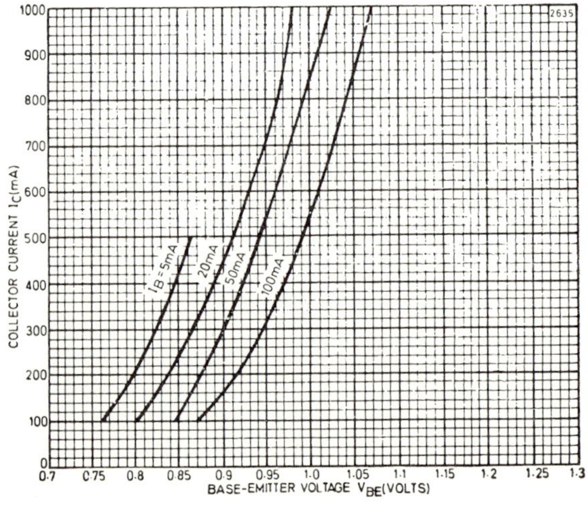

ZTX 300 (BCW 10) Series

The ZTX300 Series are general purpose n-p-n transistors electrically similar to the ZT80 Series and complementary to the ZTX500 (BCW11) Series p-n-p transistors.

RATINGS

		ZTX 300 (BCW10)	ZTX 301 (BCW12)	ZTX 302 (BCW14)	ZTX 303 (BCW16)	ZTX 304 (BCW18)	
Max. Collector-Base Voltage	V_{CBO}	25	35	35	45	70	V
Max. Collector-Emitter Voltage	V_{CEO}	25	35	35	45	70	V
Max. Emitter-Base Voltage	V_{EBO}	5	5	5	5	5	V
Max. Peak Collector Current	I_{CM}	500	500	500	500	500	mA
Max. Dissipation (All Types)	P_{tot}						
at T_{amb}=25°C		300 mW					
Derating with temp.		Linearly to zero at 125°C					
Max. Operating and Storage Temp. (All types)		−55 to 125°C					

CHARACTERISTICS (At 25°C ambient temperature unless otherwise specified)

Parameter	Symbol	ZTX 300 (BCW10)	ZTX 301 (BCW12)	ZTX 302 (BCW14)	ZTX 303 (BCW16)	ZTX 304 (BCW18)	Conditions		
Max. Collector-Base Reverse Current	I_{CBO}	0·2	0·2	0·2	0·2	0·2 μA	$V_{CB}=V_{CBOM}$	$I_E=0$	
Max. Emitter-Base Reverse Current	I_{EBO}	0·2	0·2	0·2	0·2	0·2 μA	$V_{EB}=4V$	$I_C=0$	
Min. Collector-Emitter Sustaining Voltage	$V_{CEO(sus)}$	25	35	35	45	70V	$I_C=5mA$	$I_B=0$	
Max. Collector-Emitter Saturation Voltage	$V_{CE(sat)}$	0·35	0·25	0·25	0·35	0·35V	$I_C=50mA$	$I_B=5mA$	
Base-Emitter Saturation Voltage	$V_{BE(sat)}$								
Minimum		0·65	0·65	0·65	0·65	0·65V	} $I_C=10mA$	$I_B=1mA$	
Maximum		1·0	1·0	1·0	1·0	1·0 V			
D.C. Current Gain	h_{FE}								
Minimum		—	—	20	—	—	$V_{CE}=6V$	$I_C=100\mu A$	
Minimum		50	50	100	50	50	} $V_{CE}=6V$	$I_C=10mA$	
Maximum		300	300	300	300	300			
Minimum		—	—	50	—	—	$V_{CE}=6V$	$I_C=50mA$	
Min. Gain-Bandwidth Product...	f_T	150	150	150	150	150 MHz	$V_{CE}=6V$	$I_C=10mA$	f=100MHz
Max. Output Capacitance ...	C_{ob}	6	6	6	6	6 pF	$V_{CB}=6V$	$I_E=0$	f=1MHz
Typ. Noise Figure	N	7	7	7	7	7 dB	f=1kHz	$R_S=1500\Omega$	$I_C=100\mu A$

ZTX 107 Series

These n-p-n transistors are electrically equivalent to the BC107, 108, 109 series.

RATINGS

	Symbol	ZTX107 107A 107B	ZTX108 108A 108B 108C	ZTX109 109B 109C	Unit
Max. Collector-Base Voltage	V_{CBO}	60	45	45	V
Max. Collector-Emitter Voltage	V_{CEO}	50	30	30	V
Max. Emitter-Base Voltage	V_{EBO}	5	5	5	V
Max. Peak Collector Current	I_{CM}	100	100	100	mA
Max. Dissipation (all types) at T_{amb} = 25°C	P_{tot}	300	300	300	mW
Derating with temp.		3	3	3	mW/°C
Max. Junction Temperature	T_{jm}	125	125	125	°C
Storage Temperature Range (all types)			−40 to +125		°C

CHARACTERISTICS (at 25°C ambient temperature unless otherwise specified)

Parameter	Symbol	ZTX107 Min.	ZTX107 Max.	ZTX108 Min.	ZTX108 Max.	ZTX109 Min.	ZTX109 Max.	Unit	Conditions
Collector-Base Reverse Current	I_{CBO}	—	0·015	—	—	—	—	μA	$V_{CB}=50V$, $I_E=0$
Emitter-Base Reverse Current	I_{EBO}	—	—	—	0·015	—	0·015	μA	$V_{CB}=30V$, $I_E=0$
Base-Emitter Voltage ...	V_{BE}	0·6	0·7	0·6	0·7	0·6	0·7	V	$I_C=2mA$, $V_{CE}=5V$
Collector-Emitter Saturation Voltage	$V_{CE(Sat)}$	100 Typ.		100 Typ.		100 Typ.		mV	$I_C=10mA$, $I_B=1mA$
		200 Typ.		200 Typ.		200 Typ.		mV	$I_C=100mA$, $I_B=10mA$
Collector-Emitter Knee Voltage	$V_{CE(Knee)}$	500 Typ.		500 Typ.		500 Typ.		mV	Measured at $I_C=10mA$ when I_B is adjusted such that $I_C=11mA$ at $V_{CE}=1V$.
D.C. Current Gain	h_{FE}								
ZTX107, 108, 109 ...		40	130*	40	130*	100	300*	—	$I_C=10\mu A$, $V_{CE}=5V$
		125	500	125	500	240	900	—	
ZTX107A, 108A		125	260	125	260	—	—	—	} $I_C=2mA$, $V_{CE}=5V$
ZTX 107B, 108B, 109B ...		240	500	240	500	240	500	—	
ZTX108C, 109C		—	—	450	900	450	900	—	
Gain-Bandwidth Product ...	f_T	150	—	150	—	150	—	MHz	$I_C=10mA$ $V_{CE}=5V$ f=100MHz
Output Capacitance ...	C_{ob}	—	15	—	15	—	15	pF	$V_{CB}=5V$, $I_E=0$, f=1MHz
Noise Figure	N	—	10	—	10	—	4	dB	$V_{CE}=5V$, $I_C=10\mu A$, f=1kHz $R_S=2k\Omega$

*Typical

Typical Small Signal 'h' Parameters	Symbol	All ZTX107 range	All ZTX108 range	All ZTX109 range	Unit	Conditions
Input Impedance	h_{ie}	4·8 Typ.	4·8 Typ.	7·3 Typ.	kΩ	} $V_{CE}=5V$, $I_C=2mA$, f=1kHz
Reverse Voltage Transfer Ratio	h_{re}	$4{\cdot}1\times10^{-4}$Typ.	$4{\cdot}1\times10^{-4}$Typ.	$6{\cdot}4\times10^{-4}$Typ.	—	
A.C. Current Gain	h_{fe}	125 500	125 500	240 900	—	
Output Admittance... ...	h_{oe}	30 Typ.	30 Typ.	40 Typ.	μmho	

NOTE: A, B and C variations of the basic types will be indicated by colour coding as shown below.

Basic Type	h_{FE} Category	Coloured Spot
ZTX107, 8	A	Red
ZTX107, 8, 9	B	Green
ZTX108, 9	C	Blue

ZTX 114

Low noise, high gain, n-p-n transistor electrically equivalent to the BC114 transistor.

RATINGS (Absolute maximum values)	Symbol		Unit
Collector-Base Voltage	V_{CBO}	30	volts
Collector-Emitter Voltage	V_{CEO}	25	volts
Emitter-Base Voltage	V_{EBO}	6·0	volts
Dissipation			
(at 25°C Amb. Temp.)	P_{tot}	300	mW
(at 25°C Case Temp.)		500	mW
(at 65°C Case Temp.)		300	mW
Operating and Storage Temp. Range	T	–55 to +175	°C
Lead Temperature (Max. Soldering Time—10 secs.)		260	°C

CHARACTERISTICS (At 25°C ambient temperature unless otherwise specified)

Parameter	Symbol	Min.	Typ.	Max.	Unit	Test Conditions
Collector-Emitter Saturation Voltage	$V_{CE(sat)}$	—	—	0·35	V	$I_C=10mA, I_B=0{\cdot}1mA$
Collector-Base Reverse Current	I_{CBO}	—	—	0·2	μA	$V_{CB}=5{\cdot}0V, I_E=0$
*Collector-Emitter Sustaining Voltage	$V_{CE(sus)}$	25	—	—	V	$I_C=10mA, I_B=0$
Collector-Base Breakdown Voltage	BV_{CBO}	30	—	—	V	$I_C=100\mu A, I_E=0$
Emitter-Base Breakdown Voltage	BV_{EBO}	6·0	—	—	V	$I_E=100\mu A, I_C=0$
D.C. Current Gain	h_{FE}	200	350	—		$I_C=2mA, V_{CE}=5V$
A.C. Current Gain	h_{fe}	1·0	1·3	—		$I_C=50\mu A, V_{CE}=5V, f=20MHz$
		3·0	—	—		$I_C=10mA, V_{CE}=5V, f=20MHz$
Gain-Bandwidth Product	f_T	—	350	—		$I_C=10mA, V_{CE}=5V, f=100MHz$
Noise Figure	N	—	1·5	3·0	dB	$I_C=30\mu A, V_{CE}=5V, R_S=10k\Omega$ $f=1kHz$

*Pulsed: Pulse duration 300μS, duty factor=1%.

BFS 96, 97, 98

Medium power p-n-p transistors complementary to BFS59, 60, 61.

Absolute Maximum Ratings	Symbol	BFS96	BFS97	BFS98	Unit
Collector-Base Voltage ...	V_{CBO}	–60	–60	–80	V
Collector-Emitter Voltage ...	V_{CEO}	–30	–40	–60	V
Emitter-Base Voltage ...	V_{EBO}	–5	–5	–5	V
Peak Collector Current ...	I_{CM}	1	1	1	A
Dissipation (all types)					
at T_{amb} = 25°C...	P_{tot}	500	500	500	mW
at T_{case} = 25°C...		2	2	2	W
Typical Thermal Resistance (all types) ...	θj—Amb.		300		°C/W
	θj—case		75		°C/W
Operating and Storage Temperature Range (all types)			–55 to +175		°C

DISSIPATION

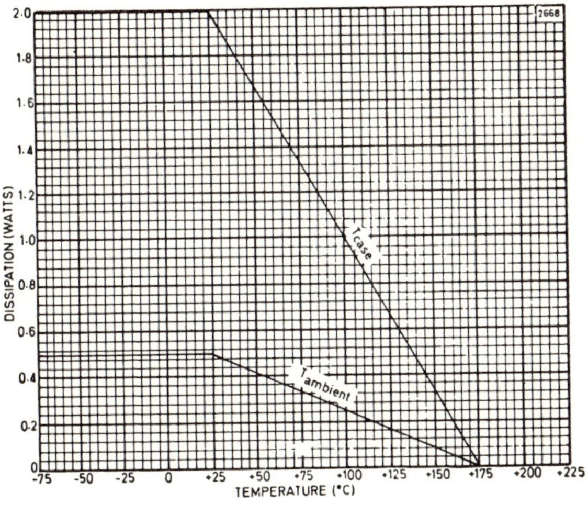

CHARACTERISTICS (at 25°C ambient temperature unless otherwise specified)

Parameter	Symbol	BFS96 Min.	BFS96 Max.	BFS97 Min.	BFS97 Max.	BFS98 Min.	BFS98 Max.	Unit	Conditions
Collector-Base Reverse Current	I_{CBO}	—	0·1	—	0·1	—	0·1	μA	$V_{CB}=V_{CBO}$, I_E–0
Emitter-Base Reverse Current...	I_{EBO}	—	0·1	—	0·1	—	0·1	μA	V_{EB}=–4V, I_C=0
Collector-Emitter Sustaining Voltage ...	$V_{CEO(sus)}$	-30	—	-40	—	-60	—	V	I_C=–10mA (pulsed)
Base-Emitter Saturation Voltage	$V_{BE(sat)}$	—	-1·1	—	-1·1	—	-1·1	V	I_C=–150mA, I_B=–15mA
Collector-Emitter Saturation Voltage ...	$V_{CE(sat)}$	—	-0·35	—	-0·25	—	-0·35	V	I_C=–150mA, I_B=–15mA
D.C. Current Gain ...	h_{FE}	40	300	100	300	40	160	—	I_C=–150mA, V_{CE}=–10V (pulsed)
Gain-Bandwidth Product ...	f_T	150	—	150	—	150	—	MHz	I_C=–50mA, V_{CE}=–10V, f–100MHz
Output Capacitance ...	C_{ob}	—	15	—	15	—	15	pF	V_{CB}=–10V, I_E=0, f=1MHz

INDEX

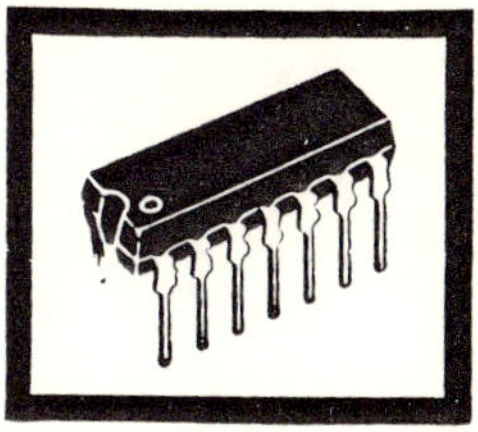

A

B

C

D

E

F

G